W9-BGA-280

Bob Coleman

ASIS International

EMERGENCY PLANNING HANDBOOK

ASIS International

EMERGENCY PLANNING HANDBOOK

INTERNATIONAL
Advancing Security Worldwide™

Second Edition

Copyright © 2003 by ASIS International

ISBN 1-887056-21-1

All rights reserved. No part of this publication may be reproduced, stored in a retrieval system, or transmitted, in any form or by any means, electronic, mechanical, photocopying, recording, or otherwise, without the prior written consent of the copyright owner.

Printed in the United States of America

10 9 8 7 6 5 4 3 2

EMERGENCY PLANNING HANDBOOK, SECOND EDITION

The contents of this handbook are the sole responsibility of the authors and do not necessarily reflect the views, experience, practices, standards, or methods of any particular company or of ASIS International as a whole.

The advice, suggestions, and recommendations given in this handbook are not meant to be adopted or implemented by readers without studying the particular facts and circumstances of their own situations. The advice of consultants or other specialists familiar with the specific facts of each case should also be considered.

The information in this handbook is based on the experience, judgment, and opinions of the authors. Neither the authors, nor their employers, the committee, or ASIS, can assume any liability for the results in a particular case of adopting, or failing to adopt, any of the recommendations made in this book.

First Edition, November 1994
Second Edition, January 2003

This handbook is an ongoing project of the ASIS International Disaster Management Council. Many thanks to the following council members, past and present, for their contributions.

Acknowledgments

Robert G. Lee, CPP
Major Author – First Edition

Special Recognition

Gregory A. Gilbert, CPP
Contributor - First Edition
Editor-in-Chief of Second Edition
New Covenant Church of Philadelphia

Lawrence L. Binkley, CPP
Wells Fargo Security Services
Contributor - First Edition

Dale A. Moul, CPP
Battelle Columbus Division
Contributor - First Edition

Aaron M. Kramer
Battelle Columbus Division
Contributor - First Edition

Allan Schwartz, CPP
Safeguards International, Inc.
Contributor - Second Edition

Werner Preining, CPP
Interpool Security, Ltd.
Contributor - Second Edition

TABLE OF CONTENTS

1.0 INTRODUCTION

1.1 PURPOSE AND SCOPE

This Emergency Planning Handbook provides guidance and direction to corporate security supervisors/managers who have emergency planning responsibilities. Its intent is to assist managers in developing sound emergency plans within their corporate planning processes.

The handbook also provides an overview of recovery planning and its relative importance to an organization. However, the overall topic of business continuity and recovery planning is expansive and is beyond the scope of the handbook. The handbook provides basic planning considerations and guidance for those individuals who need to develop a recovery plan. References providing more extensive information on recovery planning are listed in **Appendix A**.

This handbook emphasizes concepts derived from a variety of references and the experience of members of the ASIS International Disaster Management Council. The intent at this time is not to provide standards but to impart planning guidance in summary form that can be adapted to and supplemented by company procedures and policies. In order to promote understanding of terminology, a listing of selected concepts and terms is included as **Appendix B**. Considerations relating to standards are being considered by ASIS International for future publications in this arena and others.

Suggestions for improving this handbook should be sent to the Chairman, ASIS International Disaster Management Council, through the Educational Publications Manager at ASIS International Headquarters.

1.2 BACKGROUND

Many companies lack a good emergency management plan. Stephen B. Fink's survey of Fortune 500 companies, published in his book *Crisis Management: Planning for the Inevitable*, notes that while 89 percent of corporate executives believe that some crisis is inevitable, half admitted

that they have no plan in place to deal with a crisis. Of the survey respondents who believed their company had a 50-50 chance of
experiencing a crisis, 36 percent had no plan in place. Most surprisingly,
42 percent of the survey respondents whose companies had experienced a serious crisis have developed no plan for dealing with the next
one.

Every business, large or small, public or private, should have some
form of emergency management plan. An "emergency," in business
planning terms, is any situation that diverts an organization from its
usual operations or production, wasting time and/or financial resources. An emergency situation arises from a "crisis"—a turning
point—and escalates in intensity. Results of poorly managed emergencies range from negative public perception to sabotage and terrorism.

Most organizations experience crises all the time; most crises are kept
from escalating to unmanageable proportions, with or without a plan.
However, an organization's effectiveness is truly tested by the few crises
that grow into emergencies. In these situations, the implementation of a
well thought-out emergency management plan can mean the difference
between success and failure.

It is the policy and intent of most companies to provide a safe and
healthful work environment for all employees and to safeguard the
equipment and facilities used by those employees in conducting business activities. As part of this overall program, companies need to have
an established emergency management plan that provides the framework and structure to manage emergency events. Failure to do the
necessary planning could seriously impact a company's ability to minimize loss of life, loss of assets, and business downtime, should an
event occur.

A full understanding of emergency planning involves an understanding
of some of the facets involved and the specific terminology used. The
following short list shows topics that are often used interchangeably
and incorrectly. Although each is a part of the puzzle, each represents its
own discipline. It will be important to keep these individual definitions
in mind and separate when thinking about the various aspects of planning and implementation steps.

- DISASTER MANAGEMENT
- DISASTER RECOVERY
- BUSINESS RECOVERY
- BUSINESS CONTINUITY

1.3 PLANNING OVERVIEW

1.3.1 What Is an Emergency Management Plan?

An emergency management plan describes the actions to be taken by an organization to protect employees, the public, and assets from threats created by natural and man-made hazards. In developing an effective emergency management plan, managers anticipate possible threats and make all the initial decisions ahead of time, so that they can focus their time and attention on the most important actions required in the event of an emergency. The emergency management plan outlines specific steps to follow in the event of a real crisis situation (incident management), and provides specific measures for recovery after the crisis has passed.

An emergency management plan should be designed to prevent, whenever possible, any incident that might cause loss, and to control incidents that cannot be prevented so that they cause a minimum of damage. For example, hurricanes and floods can usually be predicted, and although some damage usually cannot be avoided, advance planning allows time to take actions to minimize damage. Some emergencies, such as earthquakes, provide little or no warning; and although the initial damage may not be avoidable, it may be possible to limit later losses by implementing an emergency management plan as soon as the incident occurs.

The emergency management plan should be printed in an easily read information-mapping format with job specific responsibilities and delineated emergency response procedures. Each copy of the emergency management plan should be document controlled by serialization issue to each department.

In summary, emergency planning involves deciding beforehand what to do in a real crisis situation, and what to do when the crisis occurs. An effective emergency management plan has specific instructions describing the process of getting a company to implement the plan during a crisis—and recover after the crisis has occurred.

No emergency management plan can apply to every possible situation. But a well thought-out plan can help management handle emergencies in an effective and intelligent way.

1.3.2 Why Plan?

Organizations need planning to effectively and efficiently respond to emergencies in a manner that reduces injury and loss of life; prevents or minimizes damage to property and information systems; and protects the environment. Many people believe that if their organization encounters an emergency, all is lost. This can be a self-fulfilling prophecy: People with this attitude often never get around to any type of emergency planning. Often, an emergency situation can worsen if there is no timely response or an improper response.

Organizations can and do recover from emergencies. Planning works. The World Trade Center disaster of September 11, 2001, is a prime example. Many of the businesses housed there have recovered. On the other hand, many others are permanently out of business because they lost their offices, physical resources, critical information, and/or critical personnel.

To effectively manage the consequences of an emergency situation, the organization must use its resources wisely and have a framework for response to ensure that these resources can be directed where they are most needed. Prior planning facilitates these actions. The planning process enables managers to think through, in advance, the alternatives open to them in the event of an emergency. It focuses their attention on possible consequences and exposes them to the nature of uncertainty.

1.3.3 Getting Started

The starting point for emergency planning requires:

- Defining an emergency in terms relevant to the organization doing the planning

- Establishing an organization with specific tasks to function immediately before, during, and after an emergency

- Establishing a method for utilizing resources and for obtaining additional resources during the emergency

- Providing a recognizable means for moving from normal operations into and out of the emergency mode of operation

The following activities should be done during the initial stages of the emergency planning process:

1. **Review existing plans and procedures.** Existing documents may contain much applicable information, and are particularly useful if they have been tested and validated.

2. **Gain upper management involvement and support.** Their support will be needed to get things set up properly and assure that nothing stands in the way of planning an effective emergency response. The involvement of senior management is essential at the earliest stages of the planning process. The minor demands on time and attention of top management at this stage of the program will have been well worth it when an emergency does arise.

3. **Identify facilities to be included.** All facilities, including their equipment and utilities that are intended to be included in the plan must be identified. A thorough understanding of the facilities will be essential for developing an effective plan.

4. **Conduct a vulnerability analysis.** The strengths and weaknesses of the organization should be carefully evaluated. Organizational strengths become available resources in emergency management. Organizational weaknesses should be clearly identified so that they can be addressed and procedures established to deal with them, including knowledge-base surveys of all organizational staff to determine their knowledge of responsibility in the event of any impending emergency. Prior incidents and emergencies provide a good starting point for determining vulnerability. But, be cautious. A lack of incidents does not mean there is no vulnerability.

5. **Identify resources and their relative importance.** A list of the resources (both equipment and workers) that the organization has for meeting emergency requirements must be compiled. This will give

planners a basis for developing operational concepts for the emergency management plan. Resource deficiencies (people, equipment, information systems, existing plans, etc.), that will need to be addressed or remedied must also be identified. Deficiencies in any particular resource or capability identified as necessary for emergency management or recovery should be resolved.

6. **Study the organization's demographics.** Organizational demographics include an analysis of employee population and its distribution throughout the geographical area, as well as a skills inventory to be called upon when special abilities are needed. Knowledge of employees' home locations and the ability to plot concentrations are invaluable when disaster response requires relocation of people and resources.

7. **Identify potential members of the emergency planning team.** One of the first steps should be the appointment of an individual in the organization to assume the responsibility for the plan and act as planning coordinator. The coordinator should be able to deal effectively with management and employees at all levels. A committee representing various key organizational elements should be appointed to assist and advise the coordinator, and help organize the plan. Those represented on such a committee may include specialists in human resources, information systems, communications, medical, legal, transportation, public relations, plant engineering, and security.

The development and implementation of a thorough plan is a time-consuming process and requires lead-time. Sufficient time should be allotted to complete the plan. The plan should be set up in a simplified format. It should be written by a person or group capable of gathering, understanding, logically grouping, and presenting the information as it relates to the company. Additionally, the specific responsibilities of those having a role in an emergency should be outlined so that there will be an effective response to any extraordinary situation.

1.3.4 Major Planning Considerations

Early in the planning process, it is important to select the activities to be included in the plan, based on such factors as the size and structure of the organization; its emergency response capabilities and resources (firefighting, security, medical, etc.); and the established policy or intentions regarding the organization's concept of operations for response to emergencies. For example, a small company may find it more appropriate to develop an emergency action checklist, rather than a formal emergency management plan.

Successful emergency planning deals with entire sequences of undesirable events. While in some instances all you may be able to do is warn personnel to evacuate and attempt to get outside help, other situations may necessitate a range of actions. Knowing the scope of reasonable actions beforehand will facilitate decision-making when these situations occur.

No single listing of planning considerations can be prescribed for all circumstances. The primary concern is that all important activities be properly covered. The minimum planning steps needed to establish an effective planning process are:

- Develop a thorough understanding of the company's facilities and the surrounding community

- Examine the planning environment in which the company or agency is operating

- Coordinate with key players in the planning process

The major considerations for the planning process are discussed below. These considerations form the core of the plan. (See the planning checklist provided in **Appendix C**.)

1. **Direction and Control**—A centralized emergency management structure should be used to act and make decisions in emergency situations. This may be the structure used during normal operations, or it may be modified or adapted to function only in emergencies. This organization will coordinate activities and control operating resources in an emergency.

2. **Communication**—This function deals with establishing, using, maintaining, augmenting, and providing backup for all channels of communication needed for emergency response and recovery. Effective communications require planned, established, and coordinated

response and communication procedures that everyone understands. Communications operations are more likely to work in an emergency if they are part of day-to-day operations.

3. **Alerting and Warning**—Management must receive and distribute information promptly. Key emergency response staff and employees must get timely forecasts and warnings of imminent dangers and likely events, so they can prepare for, or respond to, an emergency. Warnings must be credible and ongoing credibility must be maintained if alerts are to be believed and acted upon in a timely fashion.

4. **Facility Shutdown**—Procedures must be established for shutting down equipment and utilities during an emergency or an entire facility when evacuation is necessary. The goal is to protect company facilities, equipment, and supplies that are essential for rapid restoration of operations after a disaster. This function covers damage assessment and control, as well as assignment of responsibilities for protection of company property and information before employees leave their stations. Establishment of these procedures must also be coordinated with recovery planning as outlined in section **7.0 Recovery Operations**.

5. Evacuation—A well-planned evacuation depends on sufficient warning time to move people and resources away from a threatened area. An assortment of evacuation options should be available to the decision-maker, and tailored for different types of hazards. The plan should establish clear and detailed procedures for evacuating either part or all of the facility in an organized manner. An evacuation plan should be coordinated with the various departments in the company, with local authorities, and with neighboring companies to act as host/receiver sites. When an evacuation is ordered, it starts a chain reaction of other events that must be carried out, such as shutting down equipment. Make sure the events are written out in a clear and logical order and automated for ease of recall, if required. Also consider:

   - **Organizational and Personnel Relocation**—After an evacuation the company may be able to continue operations in a limited capacity at an alternate location. Moving employees and their families in the face of a natural disaster, if required, requires extensive planning and coordination between the business's emergency organization and the government's emergency management officials. A decision must be made early in the planning process as to what extent the business will be involved in staff relocations. These relocation procedures should also be coordinated with the planning as outlined in section **7.0 Recovery Operations**.

- **Essential Operations**—Some parts of the company's organization may be required to continue operations during an emergency. The minimum number of personnel needed must be determined, as well as the requirements for shelter and care for these individuals. These operations should also be properly coordinated with recovery planning efforts (see section **7.0 Recovery Operations**).

6. **Shelter**—Adequate shelter, protecting against both natural and man-made disasters, should be provided for employees when evacuation is not feasible. Shelter may be provided within the company's facilities or in nearby buildings. Responsibility for managing and maintaining onsite shelters should be established when appropriate. Alternatively, the plan should identify public shelter facilities that local emergency management officials have allocated to the company's employees. The shelter should have ample capacity and should be inspected for safety by appropriate government and company officials.

7. **Emergency Services**—The number of people required to perform emergency services depends on the size of the organization. Only one or two may be needed at small facilities, while there may be several dozen at larger ones. The individuals who perform emergency services are part of the emergency response team (ERT), which the company depends on to accomplish vital jobs during a emergency. The ERT should develop rapid response checklists for specific areas such as security, fire and rescue, medical, and engineering. Samples are shown in **Appendix C**.

8. **Emergency Information**—This function increases employee awareness of hazards and provides active channels for informing and advising them about what actions to take before, during, and after an emergency. Informational materials given to employees should cover emergency preparedness, safety measures, evacuation procedures, and other relevant topics. This information should be distributed, reinforced, and practiced.

9. **Media Relations**—Whenever there is a crisis at a company, the press wants to find out all the details. To avoid making a bad situation worse, companies should know how to handle the media. Depending on the organizations' needs, media relations can be a developed as a separate crisis communications plan, or incorporated in the crisis management plan.

To help prepare managers and their organizations for dealing with the media in an emergency, companies should:

- **Assess** vulnerabilities across the organization and develop an issues and vulnerability matrix to help management forecast possible problems. An issues and vulnerabilities matrix is easily developed using past problems and incidents as a guide.

- **Create** exposure alert and response processes to reduce surprises. Prepare answers to questions that are likely to arise if the company gets into the spotlight. Make sure everyone has the same response to the same questions even if the answer has to be: "No comment at this time."

- **Develop** teams of response managers and spokespeople. The goal is to create operational pairings of people, coupling the expertise in handling operational problems with skill in communications.

- **Contain** the media to a pre-determined media designated area. Establish guidelines for media access to staff.

- **Convince** top management to fully endorse and participate in emergency media relations and public affairs response processes.

- **Test** emergency management response structures and processes. The key is to be prepared to answer any and all possible questions. A company should strive to answer questions not only intelligently and coherently but also in a positive, caring, and non-threatening manner.

Procedures pertaining to media relations should also be coordinated with recovery planning (see section **7.0 Recovery Operations**).

10. **Supporting Materials**—Emergency plans and checklists should be supported by appropriate materials, including:

- **Maps**—Floor plans, street maps, and other appropriate maps. The maps can identify where staff and resources are located and help in planning strategies as an emergency unfolds.

- **Procedure Charts**—Simple organizational charts with the names, titles, addresses, and telephone numbers of key emergency personnel. The charts should show who is responsible for what tasks.

- **Call-up Lists**—Lists containing names, addresses, telephone numbers, and organizational responsibility of key emergency personnel. The lists should also have the name and telephone numbers of backup personnel.

- **Listing of Local Resources**—A listing of major sources of additional workers, equipment, and supplies.

- **Mutual Aid Agreements**—Agreements among companies and government agencies to assist one another, within defined limits, during major emergencies.

- **Glossary of Terms**—Definitions of emergency management and operational terminology to assure that everyone has the same understanding of the terminology being used.

2.0 RISK ANALYSIS

This section is intended to help managers select the most appropriate method for conducting a risk analysis. This information is not intended as a tutorial on risk analysis, but rather as guidance on developing evaluation and selection criteria.

The purpose of risk analysis is to provide the planning committee with a basis for judging the likelihood of a disaster, and the severity of its effects on the organization. The risk analysis process is a rational and orderly approach for identifying potential problems and determining their probabilities.

2.1 BASICS OF RISK ANALYSIS

Risk analysis is a procedure used to estimate potential losses that could result from various vulnerabilities and the damage from the action of certain threats. Even though an estimate is all that is possible, the answer to most, if not all, questions regarding a company's exposure can be determined by a detailed risk analysis. Risk analysis identifies not only the critical assets that must be protected, but also the environment in which these assets are located. The ultimate purpose of risk analysis is to help in the selection of cost-effective measures to reduce risks to an acceptable level through a cost effective risk management program.

Most methods of risk analysis initially require the identification and valuation of assets. From this point on, they proceed differently in computing possible loss. Most risk analysis methods or tools, however, can be categorized as either quantitative or qualitative. Quantitative risk analyses generally produce results expressed in monetary or economic terms, while qualitative risk analyses tend to use comparative expressions (such as high, medium, and low).

A thorough risk analysis provides management with well founded, reliable information on which to base decisions. Management is responsible for deciding what is acceptable, in terms of actual loss. The eventual goal of risk analysis is to strike an economic balance between the organizational impact of risk acceptance and the cost of protective measures. A properly executed risk analysis:

- Shows the current protection status of the organization

- Highlights areas where greater or lesser protection is needed

- Assembles the facts needed for the development and justification of cost-effective safeguards

- Increases hazard awareness by assessing the strengths and weaknesses of protective measures at all organizational levels, from management to operations.

The success of any risk analysis depends strongly on the role top management takes in the project. Management must support the project and express this support throughout the organization, describe with precision the purpose and scope of risk analysis, select a qualified team, and formally delegate authority. Finally, management must review the team's findings and represent the organization responsibly in accepting identified risks or taking action to minimize risks.

Personnel who are not directly involved in the analysis process must be prepared to provide information and assistance to those conducting the analysis and, in addition, to abide by any procedures and limitations of activity that may result. Management should leave no doubt that it intends to rely on the final product and base its emergency management decisions on the findings of the risk analysis team.

2.2 RISK ANALYSIS PROCESSES AND PROCEDURES

Before any corrective action—that is, action taken to minimize identified risk—can be considered, the organization must identify its risk exposure. The risk analysis involves the collection and evaluation of data concerning the physical assets, threats, vulnerabilities, and existing safeguards to determine what is at risk, and the impact or consequence if the potential threats materialize. To accomplish this, the company must:

- Identify the scope of the risk environment and determine what assets and/or resources are at risk.

- Identify all threats or potential hazards that could affect the assets in need of protection.

- Estimate the likelihood of threat occurrence.

- Determine the likelihood or the probability of damage occurring.

- Determine the impact or effect on the assets or the organization if a loss does occur.

2.2.1 Risk Identification

The primary purpose of risk identification is to make the task of risk analysis more manageable by establishing a base from which to proceed. When the risks associated with the various systems and subsystems within the organization are known, resources can be allocated more rationally. The need for risk identification rests on the premise that emergency management resources, like all other resources, are limited and therefore must be allocated wisely.

The basic considerations in identifying risks are:

- **Assets** - What does the company own, operate, lease, control, have custody of, or responsibility for, buy, sell, service, design, produce, manufacture, test, analyze, or maintain?

- **Exposure** - What, in the company's environment, could cause or contribute to damage, theft, or loss of property or other company assets, or to personal injury of company employees or others?

- **Losses** - What empirical knowledge is available to identify the frequency, magnitude, and range of past losses experienced by this and similarly located companies performing a similar service or manufacturing a similar product?

Once the answers to these questions (and to any additional questions that may result from initial inquiries) are developed, they will form the basis for the risk identification and eventually the risk evaluation, as it specifically relates to the organization.

2.2.2 Threat Identification

After identifying the activities and relationships of the organization and its assets, the next step is the process of identifying all the specific threats that must be safeguarded against. The following list of potential threats and hazards may aid in the identification process:

Natural Catastrophes

- Earthquakes, Floods, Hurricanes, Storms, Tidal Waves, Tornadoes, Typhoons, Volcanic Eruptions, Winter Storms

Industrial Disasters

- Environmental Incidents, Fires, Hazardous Material Releases, Major Accidents, Structural Collapse, Power Outages, Explosions

Civil Disturbance

- Anarchy, Anti-Company Demonstrations, Riot, Sabotage

Crime

- Arson, Assault, Bomb Threats, Bombings, Burglary, Computer Related Crime, Disorderly Conduct, Drug Offenses, Eavesdropping, Embezzlement, Espionage, Extortion, Forgery, Fraud, Gambling, Hijacking, Kidnapping, Larceny, Murder/Manslaughter, Robbery, Sex Offenses, Shootings, Shoplifting, Suspicious Packages and Letters, Trespassing

Computer Problems

- Hard Disk Crash, Data Loss or Corruption, Virus Infections, Server Crash, Website Crash, Power Loss, Firewall Failure, Hackers

Conflicts of Interest

- Bribery, Disaffection, Kickbacks, Unfair Competition

Terrorism

- Assassination, Extortion, Kidnapping, Biological and Chemical Threats

Others Threats

- Disturbed Persons, Drug and Alcohol Abuse, Drug Dealing, Medical Emergencies, Personnel Piracy, Strikes, Transportation Accidents, Industrial Espionage

Secondary Disasters

- Threats resulting from the damage caused by specific disasters. For example: an earthquake could cause a structural fire, which could, in turn, burn out circuits and cause a power failure.

2.3 CONDUCTING THE ANALYSIS

After the risk and threat identification processes, the risk analysis should concentrate on the specific risks or threats that would most seriously damage the organization. The likelihood of occurrence, along with the impact or criticality of a given disaster, should be assessed.

The relationship among the aspects of risk analysis, *risk/threat identification*, *likelihood/frequency analysis*, and *criticality*, are the fundamental elements of any safeguards method. Each aspect will be more or less significant, depending on the other aspects in any particular case. For example, if the likelihood of a given event is high, then even a relatively low level of criticality or impact to an organization will become significant. For this reason it is important that the organization take special care in collecting data concerning those assets exposed, the likelihood of occurrence, and overall impact that the specific disasters could have on the organization. It is only through such an assessment that effective countermeasures can be developed.

Various tools, manual and automated, may be used to perform a risk analysis. The most appropriate type of method depends on the size, type, and complexity of the organization performing the analysis, along with the amount of time and resources the organization is willing to allow for the process. There are numerous analysis methodologies from which to choose, and no solution is clearly the best for any given organization. A risk analysis tool should not be judged solely on the basis of how quickly it produces results, but on its ability to produce useful results with a reasonable amount of effort. The tool selected should be one that allows the user to understand how the results were reached and how they can be applied and relied upon.

2.3.1 Quantitative Versus Qualitative Approach

The risk analysis process can be time consuming and expensive. Consequently, the methodology for conducting the analysis must be chosen with care. The choice of methodology is the principle step that determines the success of the entire risk analysis process as applied to your company. Risk analysis can be performed by a quantitative or qualitative analysis or by a combination of both. Losses are derived by either mathematical (quantitative) or linguistic (qualitative) models.

Each method attempts to identify a company's exposure to threats; however, the major difference between them is the criteria used to rank the degree of an exposure. The *quantitative* approach ranks an exposure in terms of dollars and cents. Quantitative methods base a company's loss impact caused by threats on the annual loss expectancy (ALE), which is equal to the financial impact times the frequency of the occurrence (usually per year). The ALE has the advantage of translating protection needs into the financial language a business manager understands.

The *qualitative* approach, on the other hand, is more subjective, ranking an exposure by giving it a rating such as 1 through 10, for example, based on the knowledge and judgment of those doing the analysis. Although qualitative approaches to risk analysis are useful for determining an organization's exposure to risks, they do not provide dollar figures to justify an investment in security measures. However, because qualitative analysis requires less information than quantitative, the task of gathering information is less time-consuming.

Both approaches have their advantages and disadvantages, and the type of method used depends on the organization's specific needs. Quantitative risk analysis may be justified by the reasoning that cost-effective safeguards cannot be evaluated against losses unless the risks are quantified. Conversely, quantitative methods have been criticized for forcing precise estimates even in cases where there is no reliable input data. On the other hand, qualitative methodologies typically emphasize descriptions rather than calculations. Regardless of the degree of quantification, it is essential that the risk analysis be based on factual, objective data and generate an objective risk assessment. Both methods are susceptible to bias and subjectivity, and must be applied responsibly.

2.3.2 Automated Risk Analysis Tools

There are benefits and limitations to any automated risk analysis methodology. Site-specific criteria must be established before an automated methodology can be selected. Automated methodologies generally provide several advantages. First, unlike manual risk analyses that usually take months to complete, an automated methodology generally provides results in a much shorter time frame. The analysis can be carried out quickly enough to ensure that the results are not outdated by changes in the organization. Also, automated risk analysis tools are easily adaptable to operational and administrative systems of all sizes, and generally allow the user to quickly explore the results of implementing certain safeguards.

On the other hand, a major disadvantage is that there is no standard method for performing risk analysis, and there is no assurance that any particular method is complete or accurate. This can make it difficult for users to select the best risk analysis tool for their needs. The root questions in analyzing these tools must be, "What are the tools measuring, and are the results useful to my organization?"

2.4 BUSINESS IMPACT ANALYSIS

A risk analysis provides management with information on the hazards to which the facility is exposed and the potential physical losses that can occur. This information helps management determine the mitigation actions that will be implemented and the level of emergency preparedness that will be instituted.

It does not, however, provide the business impact information that management needs for recovery planning. Such information is provided by a Business Impact Analysis (BIA). The BIA is a systematic evaluation of business operations and processes that classifies and prioritizes them according to their time sensitivity and loss impact based upon downtime. The information developed in the BIA provides the foundation for developing recovery strategies and procedures and is vital to the recovery plan's overall effectiveness.

The analysis must consider what events could affect the facility and its operations, including:

- Events inside the facility.

- Events outside the facility, either in nearby buildings or businesses or in the community, that could reduce the facility's operability (e.g. loss of utilities, hazardous material spill).

- Other events that could shut down the facility or its operations (e.g. loss of a major supplier).

When completed, the BIA provides information to management on:

- What can happen—Internal and external hazard exposures and their potential consequences

- What will be affected—The business functions that will be affected, their time sensitivity (criticality), and maximum allowable downtime

- What the business impact will be—Expressed either quantitatively or qualitatively

- What major resources will be needed to re-establish specific business functions within their maximum allowable downtime.

"Maximum allowable downtime" is the period of time that a function can be discontinued without the impact on product delivery, customer service, company revenue, public credibility, or contractual arrangements going beyond a manageable and acceptable level.

Impact can be based on both tangible and intangible factors and may be expressed either quantitatively or qualitatively. Such factors include:

- Loss of revenue and other specific financial impacts

- Loss of market share

- Reduced credibility with the public and/or the financial/investment community

- Legal/regulatory impact

Using the information developed in the BIA, management can:

- Identify and prioritize those operations and processes that require recovery planning.

- Establish recovery strategies to reinstate such operations and processes within the required time frames.

- Arrange for the necessary resources to support the recovery strategies.

- Establish any financial arrangements necessary to support the recovery process.

- Identify the current level of preparedness from a recovery standpoint.

- Implement actions to mitigate the consequences and impact of an event (e.g., moving from a sole source to a dual source supply of materials).

The analysis allows decisions for recovery planning to be based upon specific information and facts, not guesswork.

3.0 EMERGENCY PREPAREDNESS PLANNING

This section covers topics that should be considered when designing an emergency management plan; that is, the planning considerations that must be addressed for a company to effectively respond to and manage an emergency event.

Emergency preparedness provides the framework around which the emergency management plan is developed. The plan defines the purpose, scope, and objectives; establishes the emergency organization structure and reporting relationships; clearly identifies what functions are included in the plan; and specifies leadership roles to provide direction and control in managing an emergency.

3.1 ORGANIZATION OF THE PLAN

The emergency management plan provides operational guidelines in a number of key areas. Key topics reflected in the following outline should be considered for inclusion:

- Introduction

  - Purpose and Objectives
  - Scope
  - References

- Emergency Organizational Structure

  - Emergency Management Team
  - Emergency Response Team

- Responsibilities

- Levels of Disaster and Emergencies

- Emergency Alerts and Notification Procedures

- Site/Area Evacuations/Shelter Areas

- Media Relations

- Control Point Operations/Supporting Materials

- Plan Activation and Responder Notification

- Plan Administration and Testing

The plan should address those measures that will be used to assure continuity of operations during the emergency and recovery after the emergency. These issues are addressed in sections **4.0 Continuity of Operations/Incident Management** and **7.0 Recovery Operations**.

The complexity of the emergency management plan will vary from company to company based on factors such as scope of the plan, company size, nature of the business, organizational structure, and identified risks. However, the key areas identified above should be an integral part of any emergency management plan.

3.2 PLAN INTRODUCTION

Purpose and Objectives

This section of the plan should clearly identify the purpose and objectives of the emergency management plan. For example, the purpose might be stated as: "To establish the organizational structure, policies, and practices to be used in managing a disaster while minimizing the impact on the company, employees, assets, and the community."

Specific objectives may be to:

- Provide a general reference for disaster and emergency management procedures, responses, and practices.

- Provide specific response plans for reasonably anticipated threats as identified by the emergency management plan.

- Establish and document disaster and emergency response actions in accordance with Federal, state, and local laws and regulations.

- Define the incident command, which will provide structure and coordination during the event.

- Develop a program for the administration and testing of the emergency management plan.

Scope

This section of the plan should define the specific limitations affecting the emergency management plan and who/what is included. For example, the geographic scope could be isolated to a single building or a multi-building site, or could include corporate-wide disaster management considerations.

References

This section of the plan should identify regulatory requirements defined by Federal, state, and local laws, as well as by company polices and procedures. Examples are:

- NFPA101 - Life Safety Code

- 29 CFR 1910 - Occupational Safety and Health Standards

- SARA TITLE III - Environmental Protection Agency Standards

- 49 CFR - Department of Transportation Standards

- Resource Conservation and Recovery Act

- Company Policies

  - Safety Standards

  - Building Closure Policy

  - Evacuation Policy

- Bank Security Desk Reference

  Federal Financial Institutions Examination Council (FFIEC) Policy on Contingency Planning. *Contingency Planning for Financial Institutions.* Disaster Requirements for Banking and Financial Institutions.

3.3 Emergency Organization Structure

Organizational approaches used in planning should be clearly described. The approach will vary from company to company based on size, type of business, and other factors.

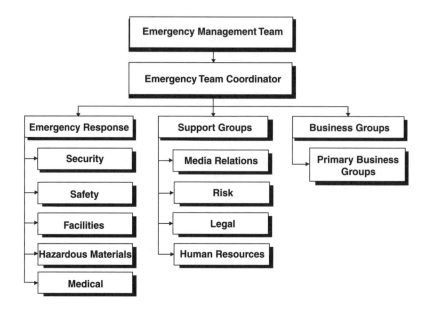

Figure 1. Emergency Management Team Structure Example

Some example organizational functions and positions that should be part of any emergency management plan are discussed below. Also provided is an example of a generic Emergency Management Team structure (**Figure 1**). This example is one of many possible approaches to emergency management organizational structure, which will vary with the size of the organization, type of business, and other factors.

Emergency Management Team

This group is responsible for developing, deploying, and testing the emergency management plan. Members of this team may also represent their organizations during an emergency. They are also responsible for the overall coordination and direction of all activities during an emergency event.

A key role of this team is to ensure the coordination of responsibilities and assignments among various organizations. The goals of the team are to reduce the risks of physical harm, stabilize situations that may arise, re-establish control, assure effective decision making, communicate accurate information during times of crisis and return to normal. Therefore, it is important that the membership reflect a good representation of emergency response groups, business groups, and support groups.

The emergency management team coordinator is responsible for providing coordination and general direction during an emergency. This individual should be a member of the emergency management team and is responsible for communicating with the top management of the organization. A backup should be designated who can assume these responsibilities if the primary is unavailable.

The on-scene incident response coordinator is responsible for the coordination of actions and activities at the scene of an incident. This individual coordinates the activities of the emergency response groups and will communicate with the emergency management team coordinator. This individual is also responsible for coordinating with response agencies or groups at the scene. In addition to the emergency response groups at the scene, the on-scene incident response coordinator also coordinates with on-scene representatives from other organizations who are part of the disaster response; for example, media relations, risk analysis, and building or business group management. A backup must be identified in case the primary is unavailable.

Other functions that must be included on the team include:

Emergency Response Groups

- Security

- Safety

- Facilities

- Hazardous Material

- Medical

Support Groups

- Media Relations

- Information Systems

- Risk

- Legal

- Human Resources

Business Groups

- Primary business groups and divisions

3.4 GROUP RESPONSIBILITIES

The plan must define responsibilities for each of the emergency response groups, business groups, and support groups activated during an emergency. Key tasks and a "game plan" to provide support during a large scale-event should be specified. The plan must designate coordination between the groups, in order to avoid confusion and duplication of effort.

Below are some of the responsibilities typically associated with certain groups. This list is not all-inclusive, and some companies may approach certain tasks differently. However, some tasks are applicable to all organizations, such as maintaining available cash, protecting or backing up key records, and keeping notification lists current.

Security

- Provides initial response—responders must be trained in, for example, first aid, cardiopulmonary resuscitation (CPR), emergency care, bloodborne pathogens, hazard awareness

- Provides crowd control

- Establishes access zones during a chemical spill

- Identifies critical security areas

- Supports building evacuation

- Provides building access control

- Provides investigative support

- Provides emergency vehicle access (public/private)

Medical

- Provides medical support at the scene of the incident

- Maintains proper levels of supplies and equipment

- Develops plan for triage at the scene

- Develops procedures for handling employees who are exposed to chemicals

- Plans for a temporary morgue, if necessary

- Establishes critical relationships with hospital, ambulance service, service providers in advance

- Maintains key contact lists for local hospitals

Facilities—provides:

- Building floor plans

- Maps of site and buildings with utility schematics showing shutdown locations, storm drains, elevators, natural gas

- Electrical power and emergency backup

- Inventory of site equipment

- Reference vendor list for available heavy equipment

Information Systems

- Maintain and provide accurate lists of employees home addresses, home phone numbers, and cellular telephone numbers.

- Quickly provide or develop lists of fax machine numbers available between key personnel.

- Provide listing of alternate email logons for personnel communications.

- Provide backups for corporate websites and information services normally on the Internet if these services are affected by the disaster.

- Provide offsite storage of critical files

- Provide offsite/hotsite for alternate operating capability

Safety/Industrial Hygiene

- Part of initial incident response

- Makes an initial assessment of hazard potential and direct defensive action to prevent injury and property damage

- Works with security to establish safe zones

- Maintains portable industrial hygiene monitoring and sample collection equipment

- Coordinates building evacuation drills and emergency notification system tests

- Identifies safety monitors in all buildings to assist with building evacuations

Hazardous Material

- Responds to emergencies involving actual or potential release of hazardous materials

- Controls, contains, and neutralizes release of hazardous materials

- Maintains necessary equipment and personal protective gear, such as fully encapsulated suits, respirators, and chemical decontamination and neutralization kits

- Maintains appropriate reference materials, i.e. material safety data sheets (MSDS)

- Ensures compliance with Occupational Safety and Health Administration (OSHA) - required training

- Monitors airborne or waterborne concentrations of released chemical

- Decontaminates all equipment, protective clothing, and tools

Media Relations

- Plans for gathering and briefing press personnel

- Works with management to develop press releases - plans for periodic updates

- Assists with internal communications

- Maintains list of television and radio stations and local newspapers

- Has necessary equipment to send or receive news and information

- Functions as spokesperson for the company

Risk (Insurance)

- Identifies actions to file insurance claims for loss or damage to company property, loss or injury to third party, and liability for bodily injury/death under Workers' Compensation Act

- Identifies insurance company points of contact

- Identifies internal financial coordinators who will assemble cost information to file claim

- Defines process for collecting loss data

- Conducts onsite survey with outside insurance adjuster

- Develops record of loss/damage, such as pictures or videotape

Legal

- Plans for backup of key documents and computer information

- Provides ongoing assessment of liabilities to the company and its personnel arising from disasters, actions in response to disasters, and actions to limit losses resulting from disasters

- Advises emergency management team and company management on recovery actions to minimize potential liabilities to employees, public, stockholders, governments, property owners, suppliers, customers, and creditors

- Participates in internal investigation of the cause of the event, actions taken, and injuries and damages

- Evaluates/investigates/defends claims, lawsuits, and other adverse actions

- Coordinates applications for government disaster relief

Human Resources

- Establishes response center with support from Information Services to receive incoming calls to company from family and friends

- Maintains employee emergency notification information—need to ensure that company has access to current employee information

- Makes notification to next of kin

- Provides support to employee families, such as benefits, transportation, inquiry response

- Establishes liaison with hospitals

- Assists families in making travel reservations

- Provides current building employee rosters to building management and response personnel with support from Information Services

- Establishes counseling assistance for employees and families

Business Operations

- Conducts regular training on emergency evacuation procedures, including evacuation routes, personnel accountability, special needs, and recognition of warning signals

- Establishes operational shutdown procedures

- Has emergency programs in place for visually or hearing impaired and handicapped employees

- Establishes telephone numbers that employees can call concerning severe weather or emergency work cancellation and audits phone lists for accuracy

- Develops program to ensure that employees receive timely communications concerning the cause of safety events and the associated corrective actions

3.5 LEVELS OF DISASTERS AND EMERGENCIES

Emergency management plans may be developed around a "worst case scenario." However, in actual practice, situations requiring emergency response may not require full deployment of the emergency response team, the on-scene incident response coordinator, and other elements associated with a full disaster response.

It may be helpful to have a classification system based on key concerns. For example, the priorities for prevention may be categorized as:

1. Injuries or deaths

2. Structural damage

3. Business interruption

4. Community impact

Prioritized classifications can be set up as relative indicators of the magnitude, severity, or potential impact of the situation:

Level 1 - Routine emergency incidents

Level 2 - Minimal business interruptions
Minimal damage
No casualties
No community impact

Level 3 - Moderate business interruptions
Moderate damage
Moderate injuries/deaths
Moderate community impact

Level 4 - Major impact in all areas

These levels may aid in planning for organizations that are developing response plans and implementation "triggers" during an emergency event. Determining the initial level of the event and the progression from one level to the next will normally be the responsibility of the emergency management team coordinator.

3.6 EMERGENCY ALERTS AND NOTIFICATIONS

3.6.1 Emergency Alerts

Companies should have in place a variety of warning systems that can alert personnel to impending threats. These systems can range from internal systems that monitor fire, intrusion, toxic gas, etc., on an ongoing basis to systems that are activated when conditions dictate, such as emergency weather radio, or the ham radio spotter network.

When emergency information is received, there should be an established procedure for quickly communicating this information to management. Emergency response personnel must be kept informed about critical developments so as to be able to activate response procedures as rapidly as possible.

Other considerations regarding emergency alerts are:

- Coordination with local city and state warning systems

- Established communication liaison with city fire and police departments

- Backup systems, including trained spotters for a variety of adverse weather situations

- Clearly defined notification procedures, complete with numbers for telephones, pagers, and mobile phones.

3.6.2 Emergency Notification Systems

Companies should have systems in place to notify employees of emergency situations and to provide direction as to what actions they should take. Companies with voice systems should have prepared wording or pre-recorded messages for different types of required actions, such as building evacuation or movement to internal tornado shelter areas.

Other considerations are:

- Stroboscopic or vibrating pager systems for the hearing or visually impaired

- Regular tests to ensure that the system works and the message is audible

- Tie-in to an emergency generator or battery backup

- Use of the emergency notification system limited to emergency events

- Decision and approval process to make emergency notification system announcement

3.7 SITE/AREA EVACUATIONS/SHELTER AREAS

Clearly established procedures should be in place for evacuating employees from company buildings or directing them to internal building tornado shelter areas. Maps should be posted in the workplace identifying evacuation routes and shelter areas. Annual drills should be conducted to test readiness. Safety monitors should be assigned to assist with evacuations. Procedures should be established to have employees gather in specific locations so that employees can be accounted for and absences noted.

An evacuation decision and approval process should be in place. Other considerations are:

- Alternative plans in case that an evacuation route is blocked

- Procedures to assist handicapped and visually or hearing impaired employees

- Coordination to confirm that a building/area has been evacuated

- Identified assembly locations at appropriate distances from the disaster scene

- Predetermined muster points for emergency response personnel

- Clearly marked evacuation routes and exits

- Process shutdown procedures

- Vital record storage evacuation plans

3.8 MEDIA RELATIONS

Media relations are a major aspect of any disaster. Disasters are newsworthy items, and any false or misleading information can make the situation worse. Therefore, it is important for the emergency planning group to designate a single public affairs officer to handle the media. This public affairs officer should be experienced in public affairs and should work in parallel with the emergency management team. All management personnel should know who this person is and how to get in touch with them during the event.

As an event unfolds, it is best to prevent overreaction, while not misrepresenting the event. Other key considerations when dealing with the media include:

- Deterring the spread of rumors

- Offering factual information

- Minimizing harmful publicity and damage control

- Confirming press credentials

3.9 SUPPORTING MATERIALS

The basic emergency management plan should be supported by appropriate appendices for reference during an emergency. The following are recommended:

- **Maps** - Building floor plans, plot plan (site plan; map of building and grounds), street maps, and other appropriate maps that can be tacked to sheets of wallboard in the Emergency Operations Center (see section **4.2.1**).

- **Organization Charts** - Simple organizational charts with the name, titles, addresses, and telephone numbers of key emergency personnel. The charts also should show which members of the emergency management team are responsible for certain actions, such as dealing with the local governments, other industries, or contractors who have emergency equipment or supplies on hand.

- **Emergency Notification Callup Lists** - Callup lists of key personnel for activating the basic plan. These lists could include names, addresses, telephone numbers, E-mail addresses, alternate websites, and organizational responsibilities for emergency operations. Alternates should be listed in case primary personnel are not available.

- **Listing of Local Resources** - A listing of major sources of additional workforce, equipment, and supplies. The data would list by company, location, equipment, and supplies available in the community. Also included would be the number of skilled workers available and their areas of expertise. The resource list should be updated at least annually.

- **Mutual Aid Agreements** - Agreements among companies and government agencies to assist one another, within defined limits, during major emergencies. The direction and control and emergency service staffs should be aware of the provisions of these agreements.

3.10 PLAN ACTIVATION AND RESPONDER NOTIFICATION

Plan Activation. Actions in responding to the emergency will begin upon confirmed notification from the security central station or other central reporting source that an event has occurred. Various functions will be activated based on type, severity, and potential impact of the incident.

Responder Notification. A relatively simple event may only require notification of the emergency response team coordinator. A higher-level event may require full deployment of the emergency response team.

The central communication location should be equipped with a current emergency notification list, which should include, for each individual who is to be notified:

- Telephone numbers

- Radio call signs

- Pager numbers

- Cellular phone numbers

- Home phone numbers

- E-mail address

- Alternate website address

Each response group should have a plan to notify additional personnel from outside their area should they be needed. The level of activity in control centers may limit the ability to make extensive personnel notifications, so alternative notification sources should be considered.

3.11 PLAN ADMINISTRATION AND TESTING

Plan administration should include a process for testing and changing the plan. As an emergency management plan is developed, it is important to indicate when the plan will be updated and by whom. This is especially critical when referring to key notification lists, which must be kept current at all times.

In a large company, a recommended approach is to identify "static" information that remains fairly constant; this would include the purpose, objectives, and organizational structure. Such information should be kept separate from "dynamic" information that is constantly changing, such as personnel lists, team membership, telephone and pager numbers, and E-mail and website addresses.

The plan should set forth the requirements for scheduled review, possibly annually, of the static portion of the plan and more frequent (monthly or quarterly) review of the dynamic portion of the plan.

Other administrative issues to be considered are:

- Distribution of the plan

- Who approves plan updates

- How plan updates are communicated to other members

- Executive approval of plan

- Updates to list of available resources

- Distribution of a checklist and key responsibilities to each group

It is extremely important that all aspects of the emergency management plan be practiced and effectively tested. The testing may start with individual groups testing their response plan or evacuation plans within their areas, progressing to limited site drills focusing on inter-group coordination, and leading to a full scale site drill involving outside response agencies, role players, etc.

Training on the plan may be accomplished by Application Service Provider (ASP) Internet connection over the company LAN or WAN. Global changes or revisions to the plan may be accomplished via this route.

The process of testing and subsequently revising the plan must:

• Keep people trained in plan specifics

• Address personnel turnover and replacement

• Address organizational changes

• Identify areas for improvement

• Maintain readiness

• Ensure that plan objectives are being met

4.0 CONTINUITY OF OPERATIONS/ INCIDENT MANAGEMENT

This section contains guidelines addressing continuity of operations during an emergency. To maintain continuity of operations, those involved in incident management must be aware of and well trained in their responsibilities and how the chain of command operates during an incident. The following topics should be considered when planning for continuity of operations or incident management:

- Continuity of management

  - Board of Directors

  - Executive succession

- Incident command structure

  - Emergency Operations Center

  - Emergency response coordination

  - Alternate headquarters

- Communications

- Information management: record keeping and safeguarding/preserving records

- Control point operations: maps, diagrams, procedure charts, emergency notification call-up lists, listing of local resources, special equipment

- Mitigation shutdown instructions

- Transportation

Many of the topics listed above will also continue into the recovery phase and should be coordinated with the recovery planning process as outlined in section **7.0 Recovery Operations**.

4.1 CONTINUITY OF MANAGEMENT

Leadership and direction are necessary elements in the successful conduct of all emergency management programs. Therefore, it is essential that business and industry emergency management plans and procedures assure that continuity of leadership is planned for and readily available in the immediate post-disaster period.

Specific measures required to develop a plan for continuity of leadership include those to:

• Assure a continuous chain of command

• Establish lines of succession for key officers and operating personnel

Board of Directors

One of the key measures needed for continuity of management is assurance that the Board of Directors can operate even though many members of the Board may be incapacitated, or the number of members needed to legally take actions (quorum) may be unavailable. Most importantly, measures need to be adopted for emergencies that would allow actions to be performed by a group not constituting a quorum.

Executive Succession

The emergency management plan must establish a management succession list that includes enough names to ensure that at least one of those on the list would be available during an emergency, regardless of when it happens. The home or other telephone numbers where key executives may be reached should be kept up-to-date and readily available.

The Boards of Directors in some corporations have established a management succession list for the corporate level as well as for key divisions or subsidiaries. The Board's decisions may take the form of resolutions. If alternatives are named to the Board who have not been formally elected by the shareholders, the plan should be reviewed by counsel to ensure that actions taken in an emergency will be lawful.

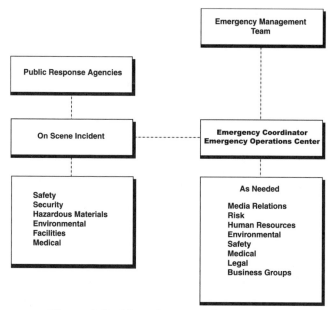

Figure 2. Incident Command Structure

4.2 INCIDENT COMMAND STRUCTURE

Figure 2 shows an example of an incident command structure that would be activated in the event of a disaster.

4.2.1 Emergency Operations Center (EOC)

The Emergency Operations Center is the location from which the emergency is managed. The emergency management team coordinator functions from this location in an emergency, with support from the various business and support groups as warranted by the circumstances.

This location functions as the nerve center during the incident and should be equipped to handle telephone, fax, radio communications, and to receive network television and computer communications of both E-mail and networked systems.

At a minimum, the EOC should be a designated room where the emergency management team could convene to manage the emergency.

Characteristics of this room include sufficient working space, appropriate emergency communications, independent and sustained power sources, and facilities where staff can eat and sleep.

The EOC will also be used during the recovery phase, and equipping it should be coordinated with the requirements of the recovery plan (see Section **7.0 Recovery Operations**).

In addition to the basic requirements listed above, several other factors should be considered when selecting a location for the EOC. First, it should be near the organization's senior management. Center operations should be separate from and not interfere with everyday activities. Operations occurring at the EOC should not be visible to visitors. Space availability needs to be considered and should be approached in terms of size of the team and the amount of equipment. Finally, backup locations and telephone numbers should be predefined and available to all emergency response groups.

The EOC and its operations can influence the outcome of any incident. There is a strong need for a well organized, easy flowing operation at the Center. In order to accomplish this, several basic needs must be met, including:

- Isolating the decision-makers from various outside conditions, such as:

  - Pressure from response personnel, the public, and media

  - Severe weather

  - Noise and confusion

- Providing a location where team members can control, locate, and identify resources

- Providing a location where team members can control overall operations

Finally, representation in the Emergency Operations Center should be predetermined based on emergency management needs as forecast during the risk analysis and documented in the emergency management plan. Additional support can be requested, as needed, based on the nature and extent of the event.

4.2.2 Emergency Response Coordination

One of the most important aspects of an effective emergency management plan is the well-coordinated response of the company's emergency response groups. As noted earlier, these usually include fire and safety, security, hazardous materials, facilities, medical, and environmental groups.

Coordination between emergency response groups through training exercises is critical to ensure an effective, well-coordinated response. Some companies have trained Emergency Response Teams to supplement the response efforts or to provide dedicated support to specific buildings. Medical Departments may have first aid/CPR/emergency care attendant trained individuals to supplement staff for large-scale medical emergencies. Coordination and communication are needed to ensure a timely, accurate response.

Some emergency response incidents do not require full deployment of the emergency management team coordinator or the on-scene incident coordinator. Therefore, it is important that the emergency management plan identify the organization that is in charge based on the type of incident; for example, Security—bomb threat, civil disturbance; Hazardous materials—chemical spill.

The emergency management plan should also include:

- A well coordinated plan for internal and external communication and notification

- Provisions for access to and storage of emergency funds

- Provisions for immediate availability of response equipment and an associated deployment plan

- Training programs that comply with local, state, and Federal laws and regulations

- Provisions for backup personnel for long term events

- A strategy for acquisition of additional response personnel; for example, cross training, contract personnel

- Provisions for mutual aid agreements with private and public sectors

4.2.3 Alternate Headquarters

It is possible that a disaster may render the main Emergency Operations Center unusable. Planning should assume this possibility, and an alternate emergency center should be established. This facility should be remotely located and furnished with the appropriate materials to keep the business flowing. Accommodations should be made for staff to eat, sleep, and work effectively.

When selecting alternate facilities, factors that may play a significant role are security, accessibility, communications, and accommodations.

This facility may be mobile or fixed: in either case, the basic essential resources include:

* Autonomous communications equipment (such as cellular phones, radios, and E-mail, if possible)

* Heat, light, and power (stand-alone generator)

* Status board or information management system

* Office/conference area

Arrangements might also be made for emergency funds to be stored at or transferred to the alternate facility. The funds should be sufficient to assure the organization's ability to get back into operation without delay.

4.3 COMMUNICATIONS

One of the most important ingredients in effectively managing an emergency event is communications. The flow of information to determine the extent of the disaster is absolutely critical so that proper response plans can be put in place.

Experts say that the media and the public often evaluate how effectively a company managed a disaster by how open and forthright the company was concerning specifics about the disaster. Therefore, it is critical that an effective communication and information flow strategy be in place.

Many disaster planners suggest that it is advisable to assume that communications will break down sometime during the event. Therefore, it is important that backup communication plans be in place. In some cases, the "backup system" may be as simple as designating a person to carry messages back and forth.

Types of communication and issues to be considered are:

- Two way radios

- Citizen's band (900 MHz)

- Hardwired telephones

- Cellular telephones

  - backup batteries

  - reception problems (e.g., in a shielded conference room)

- Predetermined list of telephone numbers and call signs for emergency communications equipment

- E-mail via laptops and the Internet or networked computers

- Alternate websites for customers and staff

- Communication strategy between emergency management and on-scene incident response personnel and the Emergency Operations Center

- Key telephone numbers for company officials, outside agencies, and customers

- Portable international communications capability, (e.g., ham radio, InMarSat)

To improve the chance of success, the communications plan should be, as much as possible, an extension of day-to-day operations rather than a complex plan to be used for disaster events only.

4.3.1. Communications Recovery Planning

A decade or two ago communication primarily meant information via voice; today data-communication has equaled and far surpassed verbal information flow. Many computer applications run automatically and unnoticed until they are disturbed, interrupted or even worse, altered by an incident. An interruption of the data-flow may cause a production process or assembly line to stop, an administrative body to idle, or an e-commerce to be unable to receive or ship orders. Depending on the type and size of the business, huge losses can occur every minute that the system is down.

In the event of a voice and/or data communication failure, it is critical to have appropriate responses ready to address this serious threat.

A communication failure will normally be the result of one of four main categories:

1. Natural causes (lightning, water damage, hurricane, tornado, earthquake, etc.)

2. Unintentional causes (human error, programming mistake, system overload)

3. Intentional causes (sabotage, vandalism, terrorism, theft of equipment or data)

4. Equipment failure (server hard disk crash, broken wire, network crash)

The following details an acceptable and simple approach for establishing a communication recovery plan.

- Conduct a *Preliminary Risk Analysis* and present it to senior management to obtain their commitment for the project. This is similar to the steps described in section **2.3 Conducting the Analysis**.

- Evaluate the communication environment (get concurrence from reputable consultant personnel for credibility)

- Get input from those who created the system

- Include realistic information relating to the systems involved

- Have a clear understanding of the types of systems that are involved

  - For example, is it still POTS[1]?

  - What type of PBX[2]?

  - T1[3] used?

  - Are there any key service contracts involved? If so, by whom?

  - Is there regular maintenance or other housekeeping done on a regular basis?

  - Who are the people that know the system best?

- Have a clear understanding of the types of service being provided (for example local carrier, long-distance carrier, 800 service, Remote Call Forwarding, etc.)

- Know the locations of equipment areas

- Know procedures for gaining access to the equipment areas

- Assure that fire protection and water damage protection is available

- Be able to identify and change critical software as required

- Understand local networks and system connectivity (LAN[4])

- Understand system protections of direct inward system access (DISA), voice mail, and remote access services

- Understand requirements and prepare standards necessary to implement a quick recovery

---

[1]POTS: Plain Old Telephone System

[2]PBX: Private Branch Exchange. It is a system that takes telephone lines from the "outside world" and makes them accessible within a certain building, office, or home.

[3]T1: "T one" is a standard 1,544 Mb/s carrier for the US, Canada, Japan, and Singapore. It carries 24 telephone lines or various broadband services from one point to another point.

[4]LAN: Local Area Network

Our primary objective should be to prevent the development of a communication disaster.

- Define and document internal standards. Plan for a successful recovery of communications within the company. Consider the possibility that key communications systems personnel may not be available as a result of the communication outage. Be prepared to implement a recovery using personnel that may not be familiar with local specialized communication issues.

Establish procedures for the following:

- Training of recovery requirement and procedures

- Testing parameters and system limitations

- Regular maintenance of all facets of the Communication Recovery

- Procedures Program

The practical application of this program should be conducted as outlined in Section **7.0 Recovery Operations**.

- Another option available in the United States is the *Government Emergency Telecommunications Service* (GETS). It is a telecommunications service provided by the Office of the Manager, National Communications System (OMNCS) that supports Federal, state, and local government, industry, and non-profit organization personnel in performing their National Security and Emergency Preparedness (NS/EP) missions. GETS provides emergency access and priority processing in the local and long distance segments of the Public Switched Network (PSN). It is intended to be used in an emergency or crisis situation during which the probability of completing a call over normal or other alternate telecommunication means has significantly decreased.

4.4 INFORMATION MANAGEMENT

Vital records—those necessary to ensure the survival of business—constitute a small part of an organization's records. It is important that vital records be given maximum protection from all types of disasters. Examples of the types of vital records that should be safeguarded are found in Section **5.1.1 Determination of Vital Records**. Preservation and protection of vital records in an emergency is essential for rapid return to normal operations. Destruction, disruption, or loss of records, even if only temporary, can significantly delay recovery operations. To assure that those records deemed essential for continuity of business are properly safeguarded, the following steps should be followed:

- Identify, in advance, priority categories of essential records.

- Label all records within the priority categories with identifiable markings. Priority of evacuations should be noted on record containers.

- Assess the vulnerability of stored records to direct and secondary damage from various disaster threats; i.e., fire, water, chemical, aftershock, vandalism, etc.

- Evaluate alternate records storage locations in light of hazard analysis.

- Make arrangements for round-trip transportation to relocate records to alternate locations, if needed.

- Identify and retain copies of the records that will be needed during the emergency operations by management or the emergency response teams.

Protection of information is further discussed in Section **5.1 Protection of Information Resources**.

4.4.1 Information Technology (IT) Recovery Planning

Even for the casual observer stepping into the New Millennium, it has become evident that Information Technology is moving at an ever increasing pace. Before entering into this new decade, big enterprises and large institutions became even more dependent on computers. Today even one-man ventures and day-to-day applications are often rendered helpless if a technology failure or computer service denial occurs. Speaking in general terms, any technology that matured over a period of some years during the 1980s and 1990s may change within months or even weeks and become obsolete. Products on the shelf can actually experience a "negative life-cycle." That is to say that once a product is out of production and on sale, it may just sit on the shelf and actually be discarded before it is ever sold. Considering the speed that technology advances, it is impossible to accurately predict how successful many products will be.

IT Recovery Planning not only deserves a fixed place in Emergency Planning, but is essential for business continuity.

Similar to physical assets, virtual property (which consists of a symbiosis of hardware and software) is at stake. The following three issues of data security need to be addressed:

1. Integrity

2. Confidentiality

3. Availability

Information Technology application may suffer from one or more of the main-threats[5] already outlined in the section **4.3 Communications**.

If your company submitted to either an ISO-certification and/or a recognized quality management/assurance program, much of the following information has already been compiled and should be on file. If the materials are on file, they will only need to be updated.

The following step-by-step instructions are general suggestions on how to establish the plan. A fine adjustment or "tuning" is normally needed, depending upon the type and scope of the business for which the particular IT recovery plan is prepared.

IT Recovery Planning Steps

- Outline the scope of the plan and the resources needed to carry it out. (For example, the newly acquired WLAN[6] or the physical environment of the data processing department. As an additional resource a consultant on encryption issues for WLAN will be needed.)

- Assure the commitment of senior management to support the plan.

- Select a competent working team out of the affected departments:
  - Communications networks
  - Database administration
  - Functional owners

---

[5]This is one of several ways to organize and catalog threats.

[6]WLAN: Wireless Local Area Network

- IT security
- Legal department (if needed)
- Physical security
- Processing operations management
- Selected system users
- Any additional department as required by the specific business

Only with the cooperation of competent persons, motivated by senior management, can such a task be successful. Team members should include professionals from the above-mentioned disciplines. Incorporating users and functional owners will encourage across-the-board acceptance of the plan versus the edict coming from security or internal audit. By experience, users and functional owners are more aware of stumbling blocks and loop-holes within the system, and are able to contribute with valuable experience.

- Identify threats and priorities

This part of the plan follows the same approach as mentioned in section **2.2 Risk Analysis Process and Procedures** and the subsequent sections **2.2.1 Risk Identification, 2.2.2 Threat Identification** and **2.3. Conducting the Analysis**.

- Establish the Total Threat Impact Report

This report is the result of the *Threat Priority* and the *Loss Impact*. Adding the fictive numbers of both *Threat Priority* and the *Loss Impact* results in the creation of the *Risk Factor*. In the example below,

Risk Factor Sheet

Identified Threats	Threat Priority	Loss Impact	Risk Factor
Fire	3	5	8
Water damage	2	5	7
Thunderstorms	2	3	5
Area Flooding	2	2	4
Human error	2	2	4
Sabotage	3	6	9

the number 2 represents the lowest, and 9 the highest possibility of occurrence and/or suffered impact.

- Identify and recommend possible safeguards

  During this phase, weaknesses in the organization are identified and *administrative, physical* and *technical* solutions sought to achieve a cost-effective level of protection. The recommendations are based upon:

  – *Avoidance*

  – *Assurance*

  – *Deterrence*

  – *Detection*

  – *Recovery*

- Prepare a Cost-Benefit Analysis

  While every phase of the prior described process is important, the most convincing part for senior management is the *Cost-Benefit Analysis*. Getting the best return on investment is something management understands clearly. Those measurements that offer a maximum protection of the identified risks at a minimum[7] of cost should be sought and recommended. Furthermore, it must be clear within the plan that the overall goals and objectives of the organization are better met through use of the suggested recommendations. Graphic depiction and spreadsheet of cost benefit analysis is easily understood and accepted by management. A picture is worth a thousand words.

- Create the Risk Analysis Report and Action Plan

  This report has two aims. It serves as a historical document that due diligence was met by those in charge to comply with their fiduciary responsibility, and to safeguard the company's recovery in an emergency and/or a crisis situation.

---

[7] Minimum cost does not equal the cheapest solution but the best monetary value for protection issues.

Example of the Cost-Benefit Analysis

Identified Threats	Risk Factor	Suggested Safeguards	Safeguard Costs	Recovery Costs
Fire	8	Automatic fire suppression system	$ 24,500	$140,000,000
Water damage	7	Integrated sensors in the data center	$ 6,000	$75,000,000
Thunderstorms	5	IT Recovery Plan	$ 25,000	$140,000,000
Area Flooding	4	IT Recovery Plan	$ 25,000	$100,000,00
Human error	4	Staff training	$ 10,000	$140,000,000
Sabotage	9	Employees	$ 10,000	$140,000,000

Some experts say that the *Action Plan* is a subset, by-product, or de-rivative of the *Risk Analysis Report*. This may be true depending on how the portions of the plan are developed. What is important is that the *Action Plan* shows clear responsibilities and necessary actions of involved departments and/or process owners, if a response is needed.

• The practical application of this program should be conducted as outlined under section **7.0 Recovery Operations** of this book.

4.5 MITIGATION SHUTDOWN INSTRUCTIONS

Procedures must be established for shutting down machinery, utilities, and processes during an emergency, or the entire facility when evacuation is necessary. Equipment that is not shut down properly could greatly increase the hazards and length of the emergency condition. Some of the types of items include heat-treat furnaces, gas generators, stills, boilers, high-pressure cylinders, and rapidly rotating flywheels. Computers and computer driven equipment also require that specific shutdown procedures be implemented, if current information is to be preserved and delicate electronic systems protected from unplanned power transients.

The emergency management plan should assign shutdown responsibilities to specific individuals who are already familiar with the process. The group of individuals should be kept as small as possible, and they must be trained in fast shutdown procedures. Shutdown of some equipment may take several hours. Simulation exercises and tagging controls will help improve speed in training. Careful shutdown procedures will greatly benefit and speed recovery.

4.6 TRANSPORTATION

During an emergency, the flow of people must run as smoothly as possible—which means minimizing panic and injury. Special transportation must be provided for any disabled or handicapped staff. All local support agencies that may be called upon for transportation should be contacted from previously prepared lists. Such agencies include area ambulance services and emergency response agencies. The organization should specifically incorporate emergency transportation into its emergency management plan. A specific person should be responsible for coordinating transportation.

The inventory of all company vehicles should be included in the emergency transportation system. A variety of vehicles should be available for all emergencies, ranging from trucks, tractors, forklifts, and wreckers to ordinary automobiles.

Trucks may be the most valuable type of vehicle during an emergency. They can serve many purposes: transporting supplies, hauling large quantities of debris, and transporting large numbers of employees and injured people.

4.7 FINANCIAL CONSIDERATIONS

Cash must be stored in an easily accessible location to assure its availability during an emergency. The amount of funds to assure continuity of operations should be determined by senior management as part of the planning process. In addition, arrangements for lines of credit and procedures for obtaining additional funds should be developed in advance. Access to cash and funds through a line of credit should be available through the Emergency Operations Center or the Alternate Headquarters. Authority for dispersing these funds should also be addressed in the emergency management plan and documented in advance of the emergency. Methods for safeguarding cash and negotiable instruments, as well as auditing procedures for accounting for this and other funds, must be established.

5.0 SPECIAL CONSIDERATIONS IN PLANNING

This section addresses additional information needs that must be considered when developing the emergency management plan, and that may be applicable to continuity of operations or recovery.

5.1 PROTECTION OF INFORMATION RESOURCES

5.1.1 Determination of Vital Records

It is important that vital records, i.e., those necessary to sustain the business, are given maximum protection from various types of disasters. The types of records a company must have in order to function varies depending upon the type of business. However, there are certain fundamental records vital to any corporate organization; for instance, the incorporation certificate, the bylaws, the stock record books, ownership and leasing documents, insurance policies, and certain financial records. More examples are listed in **Table 1**. In addition, classified and certain business sensitive documents may require special protection.

Table 1. Examples of Vital Records	
Accounts Payable	Policy Manuals
Accounts Receivable	Constitutions and Bylaws
Audits	Contracts
Bank Deposit Data	Customer Data
Capital Assets List	Debentures and Bonds
Charters and Franchises	Engineering Data
Incorporation Certificates	General Ledgers
Insurance Policies	Purchase Orders
Inventory Lists	Plans: Floor, Building, etc.

Table 1. Examples of Vital Records (con't)	
Leases	Receipts of Payment
Legal Documents	Sales Data
Licenses	Stockholders' lists
Manufacturing Process Data	Stock Transfer Books
Minutes of Directors' Meetings	Tax Records
Notes Receivable	Service Records and Manuals, Machinery
Patent Authorizations	Social Security Receipts
Payroll and Personnel Data	Special Correspondence
Pension Data	Statistical and Operation Data
Performance Appraisal Data	Telephone Directory

Depending on the type of business, additional records and information may need to be protected in order to maintain production capability. For example, a manufacturing organization may require engineering drawings and specifications, parts lists, work processes and procedures, lists of employee skills required, and similar information. Depending on the complexity of the product being produced, the task of recreating all this information would be virtually overwhelming. As another example, a banking institution may require current information on the status of depositors' accounts, accounts with other banks, loan accounts, and related banking services. Items as simple as a corporate telephone directory become critical information where no substitute is acceptable.

The emergency management plan should include a program for selecting and safeguarding vital records. The program needs to function as an administrative device for safeguarding vital information, not just for preserving existing records. This requires a systematic approach for determining what information is vital and which records contain this information. Only these records should be protected against disasters; many records may have other value to the organization, but they should not be a part of the organization's vital records protection program.

One way to identify an organization's vital records needs is to have a project team perform an assessment. The team leader might be the organization's records manager. The following sample procedure can be used by such a team to analyze an organization's vital records:

1. **Classify the organization's operations into broad functional categories**. For example:

    Finance: bill payment, account collection, and cost accounting

    Production: research, engineering, purchasing, and related activities

    Sales: inventory control and shipping activities

    General Administration: personnel, legal, tax records, public relations, and similar staff activities

2. **Determine the role of each function in an emergency.** If eliminating or reducing an activity after a disaster will restrict the organization's ability to restore some essential aspect of its operations, then that activity is vital; the information needed to maintain it is also vital and should be protected.

3. **Identify the minimum information that must be readily accessible during an emergency to assure that vital functions are performed properly.**

4. **Identify the particular records that contain vital information and the organizational entities in which they are, or should be, maintained.**

5. **Coordinate the procedures with the requirements of the recovery plan in section 7.0.**

5.1.2 Protection of Vital Records

A plan must be developed for determining the means to protect vital records. The most common method is to set up a backup system for vital records. Most systems involve storing copies of vital records at an area away from the normal office location. The dispersal of duplicate vital records lends greater assurance that the information needed to reconstruct the business after a disaster would be available in an undamaged location. While onsite storage in fireproof files, vaults, or safes may be acceptable for temporary storage of vital documents and records, a significant disaster could damage or destroy both the building and the records.

Duplicate records may be stored in a different medium than the original. A major consideration is where and for how long the duplicates will be stored. For example, some companies make duplicates on disks or microfilm, CD-ROM, streaming tape, and digital storage, all of which are sensitive to heat and water. Therefore, they must be stored in heat and water-resistant containers or locations.

Before selecting a method for protecting your organization's vital records, consider the hazards to which your site, buildings, and computer or other information storage equipment are vulnerable and the consequences of an emergency resulting from these hazards. Section **2.0 Risk Analysis** of this handbook discusses methods for hazard identification and analysis.

5.1.3 Safeguarding Vital Computer Information and Records

In addition to the protection of paper records, methods for effective protection of computer information must also be identified in the Emergency Management Plan. Much of what already has been described about selecting and protecting vital records applies to records processed by a computer. However, procedures used to protect vital data processing records must be compatible with the organization's information system design policies and computer programming concepts.

In protecting paper or microfilm vital records, it is necessary to safeguard only the record itself. In contrast, in protecting vital data processing records there are three distinctive elements:

1. Controlling of the central computer facility

2. Safeguarding the data processing media

3. Assuring information integrity and storage system compatibility

5.2 LIAISON AND INTERFACE WITH GOVERNMENT AGENCIES

The emergency management plan should designate the person or persons in an organization who will stay in contact with government agencies before, during, and after a disaster. The key here is to make sure the organization stays abreast of changing events and government activities.

Continuous communications with all government agencies (especially local) that impact your company's recovery are highly recommended. Any government agency involved with a disaster should be aware of and familiar with your policies, regulations, duties/responsibilities outlined in the emergency management plan, and the individuals designated to lead the emergency response. It is the responsibility of the company, not the government, to make sure of this. To maintain good relations, a particular person or persons (for example, the team leader) should be responsible for periodic updates and liaison with government agencies. Some examples of agencies to communicate with include local law enforcement officials, fire officials, and government representatives responsible for emergency planning.

5.3 INDUSTRIAL MUTUAL AID

Industrial mutual aid deals with resources that can be borrowed from other organizations during a crisis, as well as mutual support that can be shared with other companies. Mutual aid is a voluntary agreement among surrounding companies to assist each other during an emergency. Assistance may take the form of equipment, resources, personnel, and specialized training.

Although mutual aid can be established in a number of ways, most groups aim for the most cost-effective means. In order to do this, companies may establish specialties during training and planning activities. For example, each company may concentrate on a specific area (such as first aid, fire fighting, flood control, riots, or hazardous materials cleanup). When each company specializes in a single area, efforts in training, instruction, and actual emergency response can be reduced.

Initial steps in establishing mutual aid are:

- Get in touch with the local emergency management agency director

- Set up an informal steering committee

- Review appropriate publications

- Seek additional guidance

- Draft objectives

- Send invitations to all meetings and training sessions

- Discuss objectives

- Elect officers

- Select mutual aid coordinator

- Appoint operating committees

- Plan for utilization of resources

- Establish contractual agreements

Before integrating mutual aid into its emergency management plan, a company must have management approval, in-house preparedness, reciprocity, and reliability. It must also have detailed information concerning financial support and documentation of specific plans, responsibilities, and commitments.

6.0 TRAINING, EXERCISES, AND TESTING

The key to success of any emergency plan is training and testing. A plan cannot be expected to work properly unless it has been tested before its actual implementation during an emergency. Practicing emergency response helps assure that the response can proceed predictably in an actual emergency. By exercising the plan, problems or weaknesses in the plan and procedures can be identified, stimulating appropriate changes to the plan. The relationship of planning to training and testing is depicted in **Figure 3**.

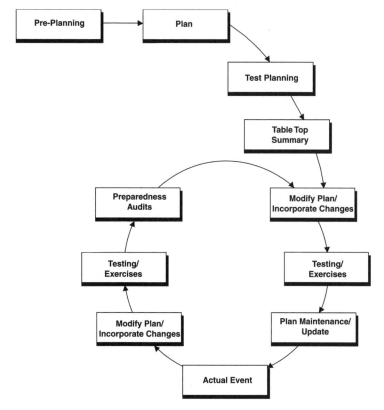

Figure 3. Training in the Planning Process

Training, tests, and exercises serve several purposes. They allow management to use and assess plans and procedures to determine whether they are actually feasible and will work under actual conditions; they assess and measure the degree to which personnel understand their emergency response functions and duties; they identify areas for improvement; they enhance coordination, communication, and proficiency among response staff; and they enhance the ability of management and staff to respond to emergencies.

Experience gained and errors committed during exercises can provide valuable insights and lessons learned that can be factored into the planning process. In addition, there may be regulatory requirements for periodic testing of emergency plans, and testing may enhance the public image of the company, since when it tests, it is acting responsibly. Inviting the fire department to a well-planned fire drill is one example of corporate responsibility in action. Inviting the news media to a test is a bit more daring but demonstrates this responsible behavior as well.

6.1 Training

All personnel should be provided with training in emergency response, commensurate with their expected level of involvement. Basically, there are three groups to which the training should be directed: employees, management, and emergency response personnel.

Employees. General employee training should assure that all employees can react automatically to warnings of an impending or imminent emergency. They should also be trained in any duties they will be expected to perform during an emergency. An important part of this training is basic understanding of the emergency management plan and how to get information and guidance during an emergency.

General training for employees should include:

- Hazards at the facility and neighboring facilities

- Warning signals and their meaning, and what response is required to the signals

- Job-specific defined responsibilities clearly describing the sequence of actions to be taken

- Sequence of actions to take in an emergency, including how to report incidents and to whom

- Identification, location, and use of emergency equipment (e.g., fire extinguishers, protective clothing, breathing equipment such as personal hoods)

- Emergency shutdown procedures

- Evacuation procedures and routes, assembly areas, and headcount procedures

Management. Leadership during an emergency is crucial to success. Therefore, more detailed training is required for those who have leadership responsibilities during an emergency. In addition to training in their duties and responsibilities for emergency response, as defined in the emergency management plan, they should understand:

- Disaster planning, response, recovery, and community linkages

- Responsibilities in the emergency management plan

- Industrial mutual aid and coordination with government agencies

- Leadership and media relations skills required for disaster management

- Special company-specific interests, such as hazardous substances

Emergency Response Personnel. Personnel who have specific response responsibilities in the emergency management plan should be trained in these specific responsibilities and the requirements of the plan and its supporting procedures. This may require specific job task analyses to determine the specific performance objectives desired. Once the performance objectives are determined, a training program should be designed to meet those objectives. For example, the emergency management team coordinator may perform a wide range of emergency duties and functions that are not related to his/her normal job. Training must assure that the emergency functions can be performed adequately. General considerations for training response personnel include:

- Threats and vulnerabilities to company facilities due to disasters

- Response procedures for events included in the plan

- Command, control, and lines of authority

- Special equipment, where it is located, and how to use it

- Equipment and systems checks (e.g., sprinkler systems, power supplies and utilities, shutdown systems.)

- Status reporting

Training administration and program development may include:

- Defined responsibility for training

- Determination of those subject to training

- Training program design (job task analysis, performance objectives, program objectives, and training methods)

- Designated instructors

- Program evaluation, quality, and revision

However simple or complex the training is, each employee and manager must know what actions they are to take in the event of a disaster, as well as what part they play in emergency response. Training should provide the background to achieve this. Further, tests, drills and exercises provide management with information as to the readiness of the company in executing the plan.

6.2 DRILLS AND EXERCISES

Drills and exercises should be preceded by training seminars or workshops, where participants are trained in their emergency responsibilities (see section **6.0 Training Exercises and Testing**). Following training in more formal settings, exercises can extend the training to provide opportunities to use existing skills and to learn new ones. Exercising the emergency management plan using simulations provides the opportunity for testing skills and knowledge to identify strengths and weaknesses; learning new skills; practicing decision-making, techniques, and communications; determining gaps in planning and procedures for management attention; and critically examining methods and procedures to stimulate revisions and modifications to the emergency management plan.

Three types of simulations that can be employed are tabletop exercises, limited scope drills, and full-scale exercises. Tabletop exercises are interactive discussions of hypothetical scenarios that take place in a small group setting. They are most useful for management team decision-making scenarios and testing the effectiveness of emergency management in responding to a host of incidents. Drills involve limited mobilization of personnel and equipment and are used specifically to evaluate and stimulate participant activity. Drills are limited in scope, providing limited testing of interaction and coordination within specifically defined groups (e.g., evacuation of a building), and they can indicate the level of participant's knowledge of required procedures. Exercises are comprehensive tests of the interaction and coordination in the emergency planning program and test the knowledge and skills of most key staff with emergency responsibilities. An exercise mobilizes personnel and equipment and uses trained personnel to control, evaluate, and simulate participant activity on a large scale. Outside agencies may also be involved.

The scope of the drill, test, or exercise is determined by what is required to ensure the learning objectives are achieved by the participants. For example, if the objective is to test the ability of senior management to make decisions as specified in the emergency management plan, a tabletop exercise would be appropriate, although the same objective could be tested during a full-scale exercise.

For simulations to be effective and achieve their objectives, results must be evaluated and reviewed. Evaluators who are not "playing" in the simulations should point out what went well and what did not. Corrective action should be initiated, based on lessons learned from the simulations. Since simulations require scenarios, guidance on formulating scenarios is provided in **Appendix D**.

Simulations also require controllers or evaluators. They must be knowledgeable in the execution of the emergency management plan and should also prepare the exercise scenario. The lead controller, who provides overall guidance to the evaluation team and coordinates with the "players", should assure that the evaluation team, as appropriate:

- Prepares player instructions (including test scope, purpose, rules and procedures)

- Prepares controller instructions and assigns controllers to specified duties

- Prepares the full scenario and player scenarios, as well as a master sequence of events

- Prepares evaluation checklists and a briefing following the exercise

These actions are highlights of the basic requirements for controlling a simulation. They may be further developed by local company experience and procedures, as appropriate.

7.0 RECOVERY OPERATIONS

This section provides an overview of the areas that must be considered and the actions that must be taken by an organization in order to rebuild after a disaster.

Although the facility may not be totally devastated by the disaster, the recovery plan must be based on the assumption that it has, and that nothing is available from the original facility to support the recovery process (the worst-case scenario). If the damage is less than total devastation, the actions detailed in the recovery plan can be scaled down to cover the requirements of the actual situation. However, it would be very difficult to "scale up" if the recovery plan was not developed on the basis of the worst-case scenario.

Care should be taken to differentiate between the emergency response procedures and the recovery plan and recovery procedures. Emergency response procedures are the actions taken before, during, and immediately after the event to mitigate the impact of the event and to stabilize the situation once the event has ceased. On the other hand, while the recovery plan and procedures may be initiated at the same time the event is ongoing, most of the plan will operate after the event has occurred. The recovery plan and procedures relate to the reinstatement of specific business functions and operations, and ultimately the return to the restored facility or a new primary facility.

The extent of the recovery plan that is actually used in a disaster will depend upon the amount of damage that has occurred, the time needed to complete repairs, and the time sensitivity of the business functions. Certain functions, such as salvage and restoration (which are normally included in the recovery plan), may also be considered emergency management functions if the damage from the event does not warrant activation of the recovery plan.

7.1 CONTENT OF THE PLAN

Recovery planning covers several major areas:

• Damage assessment to include building operability checklists

• Initial recovery of time sensitive functions

• Long-term recovery of all major functions

• Repair/reconstruction of the damaged facility or the establishment of a new primary facility

• Return to the restored facility or a new primary facility

7.2 BASIS OF THE PLAN

The plan is based upon the prior determination of :

• The time sensitivity of business functions based upon the impact of their loss. Such impact may be financial, but also may be related to public credibility, investment community perceptions, legal and regulatory requirements, etc.

• The maximum allowable time before such functions must be resumed. This is determined for individual functions or groups of related functions.

• The minimum resources needed to support the restoration of those functions in the required time frames.

7.3 TIME SENSITIVITY OF BUSINESS FUNCTIONS

The plan is based upon, and should detail, the time frames and priority in which the departments/business units and specific business functions will be re-established.

All departments/business units and their business functions have a specific time sensitivity. (This was previously known as "criticality," but because of adverse inferences that can be drawn from use of the word "critical," more references to "time sensitivity" are now being made). The time sensitivity will vary between departments/business units and also between individual business functions within the department/business unit. For example, a department may have certain functions with a time sensitivity of 24 hours (that is, they must be restored within that time frame). The other functions in that department may have a time sensitivity of five days. Another department may have functions with varying time sensitivities, starting at five days and extending up to 15 days.

The time sensitivity also equates to the maximum allowable downtime. Establishing the time sensitivity is critical to:

- Establishing the time frame for completing the initial recovery functions and the subsequent long term recovery functions.

- Prioritizing restoration of the business functions.

- Establishing the recovery strategies for restoring the business functions.

- Determining the resources that will be needed to support the restoration of the business functions and the time frame in which such resources will be needed.

7.4 RECOVERY STRATEGIES

The information on time sensitivity and the resources needed to recover the business functions will enable an organization to develop recovery strategies to:

- Restore the functions within the required time frames

- Establish alternative means of providing products or services.

Such strategies can involve:

- Specific contractual arrangements

- Use of predesignated locations

- Multiple alternatives with the actual strategy to be used only determined at the time of the incident

- Mutual aid agreements for production/manufacturing/distribution operations.

7.5 DEVELOPMENT OF THE PLAN

Development of the recovery plan involves:

- Completion of a business impact analysis

- Analysis of the alternatives available and development of the recovery strategies

- Development of procedures to activate, implement, and operate the recovery plan

- Establishment of:

  - A training and testing program.

  - Maintenance and updating procedures

7.6 ORGANIZATION OF THE PLAN

The recovery plan must detail:

- How the plan is to be administered and updated

- The assumptions used in developing both the recovery strategies and the plan

- The recovery strategies

- Damage assessment procedures

- Disaster declaration and plan activation procedures

- Notification requirements and procedures

- The recovery management team membership, roles, responsibilities, procedures, and tasks

- Recovery function teams, membership, roles, responsibilities, and procedures/tasks

- The role of the Emergency Operations Center (EOC)

- The actions to be taken by individual department/business units to re-establish their operations in temporary facilities and the procedures/tasks supporting those actions

- Details of temporary facilities/recovery sites that have been contracted for or established prior to the disaster

- Public relations/crisis communication responsibilities and procedures

- Special requirements such as:

  - Vital records

  - Management information systems recovery

  - Voice/data communication recovery

  - Insurance claim procedures and accounting

7.7 INFORMATION DATABASE

An information database should also be established and incorporated to provide the information needed to support the recovery process. Although much of this information may be available from various sources within the facility, the establishment of a central database permits:

- Standardized format

- Immediate availability in an emergency situation

- Continued availability should the original source be affected by the emergency situation

The database should include information on:

- Employees

- Customers

- Vendors/suppliers/contractors

- Equipment/assets

- Forms and supplies

- Vital records

- Computer applications/software

- Telecommunications.

7.8 RECOVERY ORGANIZATION AND MANAGEMENT

The actual structure of the recovery organization and the size and membership of the recovery management team will depend on the size and complexity of the facility and the operations covered by the plan.

The recovery organization will usually comprise:

- A recovery manager/coordinator and recovery management team

- Specific recovery function teams

- Departments, both revenue-producing and support functions that will be involved in the recovery process

7.9 RECOVERY MANAGEMENT TEAM

The recovery management team is the focal point of the recovery effort and coordinates the recovery function teams and the individual departments.

The responsibilities of the recovery management team include:

- Determining the extent of the disaster

- Relocating to the Emergency Operations Center (EOC)

- Managing the recovery and coordination between departments and between recovery teams

- Allocating the resources to the departments; if shortages occur, the team will make the allocation decisions for the departments

- Coordinating of the activities of vendors called in to assist the recovery

- Recalling employees from stand-by duties at home and temporarily assigning them to departments/teams with staffing needs

- Coordinating all requests for space, equipment, supplies, and human resource support and allocating as available

The recovery manager/coordinator is responsible for the day-to-day management of the recovery and liaison with senior management. The manager/coordinator is responsible for the disaster declaration and plan activation, implementation of the recovery strategies, and provision of general oversight and coordination of the recovery process. He/she also has total authority and accountability for the actions of the recovery management team.

The remainder of the recovery management team consists of personnel with overall responsibility for the following areas:

- Operations —Maintaining critical functions and operations

- Facilities—Assessing the damage, securing the damaged facility, repair/restoration, or finding and equipping a new primary facility

- Logistics—Providing support services and resources to the critical functions and operations

- Finance—Managing the fiscal affairs of the organization during the recovery and tracking the financial obligations incurred

- Public Relations—Controlling the distribution of information to and interacting with the media, employees, and other external agencies.

These personnel will be supported by teams/individuals with specific recovery functions in each area. Other designated internal and external specialists and administrative staff will support the recovery management team. The team size will be determined by the specific situation. It will cease to operate when management declares the situation under control and recovery complete.

The recovery management team is separate from the emergency management team. However, a recovery plan can be activated while the event is still in progress. Therefore, care should be taken to ensure that emergency management team personnel are not also members of the recovery management team or recovery function teams who would be called upon to respond, from a recovery standpoint, while the event is still in progress and the emergency response is ongoing. Such personnel can be called on to assist the recovery process when the emergency response is concluded. Failure to keep these vital functions separated can result in conflicted activity and unnecessary impediments.

The recovery management team should be operated from the Emergency Operations Center (see section **4.2.1 Emergency Operations Center**).

7.10 IMPLEMENTATION

Procedures and tasks should be developed and derived from the plan to:

- Activate and implement the recovery plan

- Manage the recovery

- Implement specific recovery functions

- Enable departments/business units to reinstate specific business function/operations at temporary locations

- Enable departments/business units to return to the restored facility or a new primary facility.

Procedures and tasks can be provided for both the recovery functions and department teams and for the individual members within each team. Major considerations in the development of the procedures and tasks are:

1. They should be sufficiently detailed to enable each person to understand their roles and responsibilities, but provide flexibility to allow changes if the circumstances of the emergency are not exactly the same as those envisioned when the plan was developed.

2. Procedures and tasks for the reinstatement of automated functions or those with substantial computer support (e.g., management information systems, telecommunications networks) may be more complex than those for other business functions.

3. Departments/business units must differentiate between their normal operating procedures and the recovery procedures and tasks. It is normally assumed that the department/business unit will follow its normal operating procedures when it is reinstated in the temporary facilities. Any expected changes to the normal operating procedures should be noted in the department's/business unit's standard operating procedures. The department's/business unit's procedures for implementing the recovery plan should detail only those procedures and tasks that are necessary to move to and setup in a temporary facility and subsequently to return to the restored facility or a new primary facility.

4. Wherever possible, titles or positions (rather than a person's name) should be used to identify the procedures and tasks to be completed. This is to facilitate maintenance of the plan and reduce the changes needed if an individual leaves or moves to another position.

7.11 TRAINING, EXERCISING, AND TESTING

Procedures should be developed and implemented to:

- Provide training to both existing and new employees. Be sure to include critical on-site contractors

- Exercise and test the plan on a regular basis (at least annually and preferably more frequently)

The exercises/tests should be designed to:

- Confirm that the information, recovery procedures, and recovery strategies are appropriate

- Discover possible omissions

- Confirm that personnel are aware of their assigned responsibilities and are capable of performing the tasks required of them

- Verify the communications back-up systems

- Verify the viability of alternate sites for production, computer operations, and the emergency operations center

- Verify that the records, information, and other items designated to be stored off-site are in place and up to date

Types of exercises/tests can vary from tabletop reviews to full scale simulations incorporating transfer to backup/alternate sites.

Section **6.0 Training, Exercises, and Testing** provides information on developing training, exercise, and testing programs.

7.12 MAINTAINING AND UPDATING

A recovery plan is a dynamic document and must be kept up-to-date and consistent with an organization's operations. In addition to pre-defined periodic reviews, the plan should specify the types of changes within the company that need to trigger a review and update of the plan. Examples include: the addition of a new site, changes in the physical site layout or manufacturing processes employed, new suppliers/vendors, added product lines, etc.

APPENDICES

American Institute of Chemical Engineers. *Guidelines for Hazard Evaluation Procedures, Second Edition with Worked Examples.* New York, NY. 1992.

American Red Cross. *Business and Industry Guide: Preparing Your Business for the Unthinkable.* American Red Cross, www.redcross.org.

Armenante, Piero M. *Contingency Planning for Industrial Emergencies.* New York: John Wiley and Sons, 1991.

Auf Der Heide, Erik, MD, FACEP. *Disaster Response.* St. Louis: C.V. Mosby Co., 1989.

Barton, Laurence. *Crisis in Organizations Managing and Communicating in the Heat of Chaos.* Cincinnati: SouthWestern Publishing Co., 1993.

Bell, Judy K. *Disaster Survival Planning: A Practical Guide for Businesses, revised edition.* Disaster Survival Planning, Inc. Port Hueneme, CA, 2000.

Berman, Alan. *Down but Not Out. A Guide to Testing Your Disaster Recovery and Contingency Plan.* Boston: Auerbach Publishers, Inc., 1988.

Broder, James F. *Risk Analysis and the Security Survey, Second edition.* Boston: Butterworth-Heinemann, 2000.

Carlton, Yvonne A. "The Plan's the Thing (Disaster Management Planning)." *Security Management,* 31:8, August 1987, pp.71–72.

Chemical Manufacturers Association. *Crisis Management Planning for the Chemical Industry.* CMA, 1991.

Chemical Manufacturers Association. *Site Emergency Response Planning Handbook.* CMA, 1992.

Chemical Manufacturers Association. *Community Emergency Response Exercise Handbook.* CMA, 1992.

Childs, Donna R., and Dietrich, Stefan. Contingency Planning and Disaster Recovery: A Small Business Guide. New York: John Wiley & Sons 2002.

Cross, Richard F., *Bank Security Desk Reference. Contingency Planning for Financial Institutions, Revised edition.* Arlington, VA: A.S. Pratt & Sons Group. 2002. (Chapter 6)

Disaster Recovery Journal, Monthly Publication, St. Louis. www.drj.com.

Emergency Management Institute. *Risk Analysis.* Federal Emergency Management Agency, Washington, D.C. SM 305, 1990.

Federal Emergency Management Agency. *Comprehensive Earthquake Preparedness Planning Guidance,* Washington, D.C. 1985.

Federal Emergency Management Agency. *Disaster Mitigation Guide for Business and Industry.* FEMA 190, Washington, D.C. 1990.

Federal Emergency Management Agency. *Emergency Management Guide for Business and Industry.* Washington, D.C. 1993. (www.fema.gov/pdf/library/bizindst.pdf)

Federal Emergency Management Agency, Emergency Management Institute. *Exercise Design Course: Instructor Guide.* SM 170. U.S. Government Printing Office, Washington, D.C. 1989.

Federal Emergency Management Agency, Emergency Management Institute. *Exercise Design Course: Student Workbook.* SM 170.1. U.S. Government Printing Office, Washington, D.C. 1989.

Federal Emergency Management Agency, Emergency Management Institute. *Exercise Design Course: Guide to Emergency Management Exercise.* SM 170.2. U.S. Government Printing Office, Washington, D.C. 1989.

Federal Emergency Management Agency. *The Emergency Program Manager.* HS-1. Washington, D.C. February 1989.

Fink, Steven. *Crisis Management: Planning for the Inevitable.* Lincoln, NE: Universe.com, 2000.

Gigliotti, Richard, et al. *Emergency Planning For Maximum Protection.* Boston: Butterworth-Heinemann, 1991.

Hayen, Jr., John C. *Emergency Management Concepts for Industry and Business*. West Nyack, NY: Todd Publications, 1988.

Healy, Richard. *Emergency and Disaster Planning*. New York: John Wiley & Sons, 1969.

Kelly, Robert B. *Industrial Emergency Preparedness*. New York: John Wiley & Sons, 1989.

Lagadec, Patrick. *Preventing Chaos in a Crisis*. London: McGraw Hill, 1993.

Lykes, Richard S. *Are You Ready for Disaster? A Corporate Guide for Preparedness and Response*. Manufacturers' Alliance for Productivity and Innovation, Washington, D.C., 1990.

Mitroff, Ian I., and Pearson, Christine M. *Crisis Management—A Diagnostic Guide for Improving Your Organizations Crisis Preparedness*. San Francisco: Jossey Bass Inc., 1993.

Mitroff, Ian I., and Anagnos, Gus. M. *Managing Crisis Before They Happen: What Every Executive Needs to Know About Crisis Management*. New York: AMACOM, 2000.

Murphy, Jean H. "Taking the Disaster Out of Recovery." *Security Management*, 35:8, August 1991, pp.60–66.

National Fire Protection Association, NFPA 11600. *Standard for Disaster Emergency Management*, Quincy, MA. 2000.

U.S. Nuclear Regulatory Commission. *Functional Criteria for Emergency Response Facilities*. NUREG-0696. Washington, D.C. 1981.

Nudell, Mayer, et al. *The Handbook for Effective Emergency and Crisis Management*. Lexington, MA: Lexington Books, 1988.

Occupational Safety and Health Administration. *How to Plan for Workplace Emergencies*. OSHA-3088. Washington, D.C. 2001.

Occupational Safety and Health Administration. *Principal Emergency Response* and Preparedness Requirements in OSHA Standards and *Guidance for Safety and Health Programs*. OSHA-3122. Washington, D.C. March 1992.

Pauchant, Thierry C. and Mitroff, Ian I. *Transforming the Crisis Prone Organization.* San Francisco: Jossey Bass, 1992.

Post, R.S., et al. *Security Manager's Desk Reference.* Stoneham, MA: Butterworth, 1986.

Rothstein, Philip Jan, Editor. *Disaster Recovery Testing, Exercising Your Contingency Plan.* Ossining, NY: Rothstein Associates, Inc., 1994.

Salvage Special Interest Group. *Salvage Briefing.* Survive. London, 1992.

Tierney, Kathleen J.; Perry, Ronald W.; Lindell, Michael K. *Facing the Unexpected: Disaster Preparedness and Response in the United States.* Joseph Henry Press, 2001.

Viera, Diane L. "A Model of Disaster Management." *Security Management,* 35:8, August 1991, pp.68–77.

Washington State Office of Emergency Services. *Business Resumption Planning Guidelines.* July 1993.

Webber, Robert. "Are You Prepared to Handle an Emergency?" *Security Management,* 29:11, November 1985, pp.20–23.

Williams, Timothy, editor. *Protection of Assets Manual.* Los Angeles: POA Publishing, LLC.

Assets: Products, processes, and/or personnel that are critical to the organization's operations. Identifying assets is the central feature of the risk analysis process. The risk analysis methodology should allow the analyst to define what is to be protected and their value. Assets may be categorized as tangible and intangible. Examples include: facilities, hardware, software, supplies, documentation, personnel, reputation, and morale.

Annual Loss Exposure: The projected loss (in dollars) that one can expect to lose in a year as a result of emergencies.

Business Recovery Planning: The process of developing the capability to offset the effects of business disruption. The process involves arranging alternatives for critical business functions and planning for business or service survival.

Consequence/Outcome: The undesirable result of a threat's action against the asset, which results in measurable loss to the organization.

Crisis: A wide variety of events that cause significant disruption to the normal activities of an organization as a whole.

Crisis Management: A planned, systematic response that permits an organization to continue making its products or providing its services during an emergency. It allows the organization to capitalize on the expertise of personnel from various disciplines who plan for and manage the situation.

Emergency Preparedness: The planning considerations that must be in place for a company to effectively respond to and manage an emergency event.

Likelihood of Occurrence: A measure of the probability of a loss-causing event.

Risk: The potential for causing losses due to the presence of a threat and vulnerability. A risk is derived from the analysis of a threat and corresponding vulnerabilities along with the probability of their interaction.

Risk Analysis: A procedure used to estimate potential losses that could result from various vulnerabilities and the damage from the action of certain threats. Risk analysis identifies both the critical assets that must be protected and the environment in which these assets are located.

Risk Exposure: The disclosure of high probability vulnerabilities.

Safeguards: Physical controls, mechanisms, policies, and procedures designed to protect assets from threats.

Threat: A person, thing, event, or idea that poses some danger to an asset. The actions of a threat may compromise the confidentiality, integrity, or availability of an asset by exploiting vulnerabilities or weaknesses in the safeguards system.

Vulnerabilities: Weaknesses in the safeguards system, or the absence of safeguards. Vulnerabilities can be clearly associated with threats: for example, the threat of fire is associated with the vulnerability of inadequate fire protection, and the threat of unauthorized access can be linked to inadequate access controls.

The following checklist is provided as a guide to assure that relevant considerations are identified in the emergency management planning process.

Use the blank space provided to record the current status of the plan or checklists:

Yes = Complete

No = Requires Action

N/A = Not Applicable

(Unknown is not an acceptable answer)

The provisions listed below correspond to the discussion in section **1.3.4 Major Planning Considerations,** and are suggested for consideration in developing a plan/planning checklist. The checklist can be used prior to developing the emergency management plan to review and evaluate organizational preparedness status and to determine planning voids and weaknesses.

Direction and Control

Does your plan or checklist have provisions for:

_____ Indicating who is in charge for each emergency or disaster situation and citing the location of the Emergency Operations Center (EOC) or on-the-scene command post from which direction and control will emanate?

_____ Determining the need to evacuate the facility or site or when to issue evacuation orders?

_____ Identifying the individual responsible for issuing evacuation orders and how they will be announced?

_____ Identifying an alternate EOC site to serve as a backup if the primary EOC is not able to function?

_____ Identifying the personnel assigned to the EOC for emergency operations?

_____ Identifying lines of succession to assure continuous leadership, authority, and responsibility in key positions?

_____ Providing for logistical support for food, water, lighting, fuel, etc., for the emergency response force?

_____ Timely activation and staffing of emergency response teams and/or personnel?

_____ Assigning operational and administrative support for emergency response activities?

_____ Clear and concise summary of emergency functions, direction and control relationships, and support communications system?

_____ Ensuring that EOC staff members can be recalled on short notice?

_____ Describing EOC functions, layout, concept of operations, duties of staff, use of displays, and process to bring the EOC to full readiness on a 24-hour basis?

_____ Protecting resources (essential personnel and equipment) during disaster situations?

_____ Implementing resource controls?

_____ Safeguarding essential records?

_____ Disaster effects monitoring and reporting capability?

_____ Central coordinating point(s) for receiving, analyzing, reporting, and retaining (events log) disaster related information (property damage, fire status) for EOC staff and/or response teams?

_____ The EOC staff to acknowledge/authenticate reports?

Communications

Does your plan or checklist have provisions for:

_____ Primary and backup radio communication, with gas generators or extra batteries (fixed and mobile, as available)?

_____ Describing the methods of communications between the EOC and response teams, dispersed company/plant operating locations, adjacent firms, and local government emergency services (fire, police, etc.)?

_____ Two-way radio communication requirements for emergency response forces, if available?

_____ Assuring that the response team members (and their backups) assigned to communications tasks understand communications terminology, and know where to obtain communications equipment and how to operate it effectively?

_____ Recalling communications staff members on short notice?

_____ Obtaining additional telephone services during emergencies?

_____ Listing key telephone numbers for industry emergency assistance organizations?

Alerting and Warning

Does your plan or checklist have provisions for:

_____ Receiving warning from the weather service or local government when hazardous situations threaten the facility?

_____ Warning the employees in the event of a disaster?

_____ Describing the warning system (type of devices, e.g., alarms, paging systems, detectors, word-of-mouth) used to alert the workers?

_____ Alternate means of warning to back up the primary system?

_____ Defining the responsibilities of departments or personnel and describing activation procedures?

_____ Warning local government and nearby establishments of onsite disasters that might spread to areas outside the facility?

_____ Requesting emergency assistance from local government (fire, police, medical, etc.)?

_____ Differentiating warning signals that identify specific threats or require specific response actions?

_____ Warning any hearing impaired and non-English-speaking workers?

_____ A 24-hour warning point to alert key officials and to simultaneously activate all warning devices?

_____ Call-up procedures to notify key officials and/or request offsite assistance in the event of an emergency?

_____ Routine checks of the warning system to assure that it is functioning properly?

Facility Shutdown

Does your plan or checklist have provisions for:

_____ Indicating under what conditions shutdown must occur or be considered?

_____ Identifying who will make the decision to shut down equipment, utilities, or the facility?

_____ Specifying who is responsible for carrying out shutdown? Assigning specific roles for equipment and utility (e.g., gas and water) shutoffs, and for checking automatic shutoffs (and for doing it manually if the automatic system fails)? Identifying who is to be equipment shutoff backup? Requiring report of shutdown completion to EOC?

_____ Establishing prearranged order or signal to initiate shutdown procedures appropriate for the impending hazard?

_____ A complete checklist for emergency shutdown?

_____ Diagrams to show where to turn everything off?

_____ Posting shutdown instructions on or near control panels, valves, switches, and operating mechanisms of each piece of major equipment?

_____ Instructing and training personnel to implement the emergency shutdown procedures?

_____ Designating personnel to close doors and windows, tie down loose equipment, move equipment and supplies to shelter area, and barricade windows and doors?

_____ Assigning personnel to stand by firefighting hoses and equipment to be ready to extinguish fires?

_____ Identifying and protecting valuable and sensitive tools, instruments, machinery, and materials?

_____ Protecting equipment and material stored outside by banding tiedown, moving critical or valuable items to inside storage, or moving mobile equipment to high ground or to protected sides of the buildings, as circumstance requires and time allows?

_____ Establishing damage assessment and control techniques to minimize property loss during a disaster?

_____ Testing shutdown procedures for utility services and equipment by department managers?

Evacuation

Does your plan or checklist have provisions for:

_____ Describing the conditions under which evacuation would be ordered?

_____ Developing evacuation procedures, with appropriate options for the various hazards, that avoid potential secondary hazards (i.e., live high voltage wires that could fall; fuel lines that could be ruptured by earthquake explosion; fire damage; etc.)?

_____ Identifying the individual responsible for ordering an evacuation and establishing lines of succession for carrying out evacuation functions?

_____ Indicating under what conditions it would be safe to complete facility shutdown before ordering general evacuation?

_____ Describing the alerting and communication systems for signaling impending or immediate evacuation for each type of evacuation your facility may require?

_____ Procedures for search and rescue teams, if evacuation alarms are inoperative?

_____ Maps indicating evacuation routes from buildings and the facility site?

_____ Clearly marked evacuation routes throughout company facilities, with two exit options (and fire escapes where needed) for every employee?

_____ Safety lighting (to ensure adequate light for evacuation during a power outage) in stairwells and corridors?

_____ Assuring that all personnel know the evacuation routes, routines, and check-in procedures for both area and site evacuations?

_____ Helping any handicapped employees to evacuate?

_____ Special attention to ensure that any non-English-speaking employees understand warning signals and know where and how to evacuate the work place?

_____ Identifying public or company provided safe reassembly areas that will not leave evacuees exposed to adverse weather conditions—below freezing temperatures, driving rains, etc.—or to radiological hazards following a nuclear incident or attack?

_____ Assigning responsibility in an evacuation to a rear guard to ensure that all personnel get clear?

_____ An organized head-count to ensure that all facility occupants have exited?

_____ A system for identifying missing persons?

_____ Ensuring that vital records are evacuated?

_____ Identifying critical equipment to be evacuated and explaining how and by whom it will be moved?

_____ A facility status report to specified company and civil authorities from the responsible onsite person following a site evacuation?

_____ Periodic evacuation drills for all facilities?

_____ Designating responsible staff members (by name and title) to maintain and update the evacuation plan on a standby basis?

Shelter

Does your plan or checklist have provisions for:

_____ Identifying existing shelter space in company facilities?

_____ Orderly movement to on-site shelter, with a general traffic pattern and ready-made directional signs?

_____ Assigning corridor, floor, and building wardens to assist employee movement?

_____ Crisis stocking of food, water, medical supplies, and other necessities for fallout shelter stay (for on-site company shelters only)?

_____ Designating shelter managers and support staff?

_____ Obtaining radiation measuring devices from local emergency management officials?

_____ Arranging training for shelter managers and radiological monitors from local and state emergency management officials?

_____ Receiving and registering additional people from nearby areas, in close coordination with government officials, if company facilities have been included in the local in-place fallout shelter inventory?

_____ Coordinating with local authorities to identify shelter locations assigned to company employees outside the facility in accordance with the local in-place shelter allocation?

_____ Printed instructions advising employees of shelter locations and routes to get there, either within the facility or nearby?

_____ Identifying the individual responsible for maintaining onsite shelters?

_____ Assuring that key workers required to continue essential operations are provided blast shelter in or near the workplace?

_____ Coordinating all key worker shelter needs with the local government?

_____ Determining when occupants can be released from shelter?

Emergency Services

Does your plan or checklist have provisions for:

_____ **General Services** (may not be applicable to every emergency service)

_____ Maintaining current notification/call-up rosters for each emergency response team (ERT)?

_____ Advising personnel of specific risks associated with handling hazardous materials and of the best means to protect themselves?

_____ Obtaining appropriate equipment, instruments, antidotes, and protective clothing for ERT members to perform emergency tasks in a hazardous material, chemical, or radiological environment?

_____ Assuring that ERT members understand how and when to use response equipment, instruments, antidotes, and protective clothing?

_____ Establishing a routine for team members to check for contamination and to dispose of contaminated clothing?

_____ Standard operating procedures for each response team, describing how the team will accomplish its assigned tasks and how it will deal with the various agencies?

_____ Entering into mutual aid agreements with other private sector companies, state and local government service agencies, and volunteer agencies?

_____ A plot plan (site plan, map of buildings and grounds), including utility shutoff locations; water hydrants and mains; storm drains and sewer lines, fences, gates; natural gas, chemical pipelines; name of each building; and street names and street number directions?

_____ A building plan (floor plan for each building), including room layout, indicating the materials to be typically found in each room or area, with notes on quantities and storage containers?

_____ Supplying copies of the organization's plot and building plans to local fire and police departments?

_____ Handling inquiries and informing families on the status of employees separated from them, especially if injured or missing, due to a disaster event?

_____ Logistical support during emergency operations?

_____ Reporting the appropriate information (casualties, damage assessment, evacuation status, etc.) to the EOC during emergency operations?

_____ Direction and control of ERT personnel during operations?

_____ Designating a representative for each ERT to report to the EOC to advise decision makers, to coordinate the team response?

_____ Recovery operations during disaster events?

Specific Services

Security

_____ Traffic control during an emergency

_____ Assisting movement to shelter or to evacuate the facility

_____ Security for critical resources

_____ Keeping order in emergency shelters

_____ Protecting company property in damaged area

_____ Evacuating disaster areas during emergency operations

_____ Training in sabotage prevention for security force

Fire and Rescue

_____ Deploying fire/rescue teams and equipment in the event of an emergency

_____ Storing fire control equipment where it will be accessible despite direct hazard effects (earthquake, fires, etc.)

_____ Assuring that team members know how to operate rescue equipment

_____ Fire protection in emergency shelters

_____ Advising decision makers about the risks associated with hazardous materials

_____ Rescuing injured people during emergency operations

_____ Alerting all emergency services of the dangers associated with technological hazards and fire during emergency operations

_____ Training in radiological monitoring

Health/Medical

_____ Selecting and setting up emergency casualty station for screening casualties, administering first aid, initiating identification and casualty records, and arranging transportation to medical facilities if necessary

_____ Obtaining emergency medical support during an emergency

_____ Maintaining an adequate inventory of medical supplies for emergency use

_____ Emergency procedures for exposure to on-site chemicals and for dealing with the injured who may also be contaminated

_____ First aid training for personnel assigned to supplement medical staff

_____ Health/medical care at any facility shelter

_____ Information programs to ensure good health under shelter conditions

Engineering

_____ Establishing and testing shutdown procedures

_____ Precautions, as necessary, to protect equipment during shutdowns and to preserve it over extended periods of nonuse

_____ Maintaining drawings showing locations of utility key valves, switches, feedlines, and hazardous areas

_____ Backup electrical power to the EOC and essential production lines

_____ Preparing and maintaining a resource list identifying source, location, and availability of earthmoving equipment, dump trucks, fuel, etc., to support disaster response recovery operations

_____ Damage assessment reports

_____ Restoring utilities to critical and essential facilities

_____ Post-disaster repairs and restoration of facility and services

_____ Sanitation services for emergency facilities

_____ Maintaining adequate water supply after shutdown for drinking, firefighting, decontamination, and sanitation

Emergency Information

Does your plan or checklist have provisions for:

_____ Assigning responsibility to assure that all employees understand the warning signals, receive general instructions on what to do in an emergency, and know where to go and how to get to their shelter areas and/or disaster stations?

_____ Preparing emergency employee guidance material based on all hazards affecting the company?

_____ Distributing emergency information materials to employees?

_____ Disseminating emergency information and instruction routes, etc., on bulletin boards and other prominent areas of the building?

_____ Providing special instructions to any key workers expected to continue operations on their roles, including information about provisions made for their safety and that of their families?

_____ Including emergency response activities on the agenda of regularly scheduled meetings for supervisory staffs?

_____ Ensuring supervisors and foremen meet regularly with their staffs to discuss the provisions of the emergency management plan?

_____ Providing routine briefings for all employees when they first enter the company to acquaint them with the emergency management plan and the response roles they will be expected to assume?

_____ Scheduling general training in safety measures for all employees and specific response action training for all response team members on a regular basis?

_____ Designating an information office to act as official point of contact during an emergency?

_____ Assigning the responsibility of spokesperson for all contacts with the news media?

_____ Providing an established procedure for authenticating all sources of information received and verifying such information for accuracy?

_____ Providing rumor control?

Administration and Logistics

Does your plan or checklist have provisions for:

_____ Assuring review and written concurrence from all company departments assigned emergency responsibilities?

_____ Assuring approval and promulgation by the chief executive of the company?

_____ Specifying the approval date?

_____ Identifying the office or individual (by job title) who is responsible for maintaining (review/update) the plan and for ensuring that necessary changes and revisions are prepared, coordinated, published, and distributed?

_____ Updating, as necessary, based on deficiencies identified through drills and exercises, changes in organizational structure, technological changes, etc.?

_____ Developing and maintaining a resource inventory listing that includes source and quality? (This listing should include lighting, first aid, medical, firefighting, and other basic emergency response support equipment.)

_____ Statements identifying additional emergency resource requirements for personnel, equipment, and supplies?

_____ Readily locating specific subjects in the plan or checklist through a table of contents and, if feasible, an index?

_____ Training response staff and specialized teams to carry out emergency functions?

_____ Reviewing those portions of the plan or checklist actually implemented in an emergency event or in an exercise to determine whether revisions can be made to improve disaster response and recovery operations?

Recovery Planning

Does your plan have provision for:

_____ Identification of the time sensitivity of business functions and their maximum allowable downtime

_____ Identification of critical business units and support units

_____ Use of a Business Impact Analysis, including periodic review

_____ A policy statement/mission statement/charter issued and signed by a senior executive

_____ Re-establishing time sensitive functions within their maximum allowable downtime

_____ Establishing when, where, and how these operations will be continued

_____ Identifying the employees who will continue the required operations

_____ Identifying and designating lodging facilities for employees, where necessary

_____ Arranging transportation for employees to alternate/backup sites or other locations that will be used to maintain services/ product delivery

_____ Supplying employees with food, water, and other essential needs

_____ Establishing and equipping an Emergency Operations Center (EOC)

_____ Consigning resources, skilled work force, equipment, and material to backup/alternate sites or other locations to be used to maintain services/products delivery

_____ Informing employees of the organization's recovery plans, their roles and responsibilities, and the resources that will be provided

_____ Establishing a recovery management team

_____ Establishing specific recovery function teams and/or individuals with specific recovery functions

_____ Detailing specific procedures and tasks for both department and recovery function teams

_____ Establishing notification procedures

_____ Clearly defined procedure for declaring a disaster and activating the plan

_____ Creation of a recovery information database

_____ Maintaining critical supplies off site if they cannot be obtained with the required time frames to support the recovery

_____ Establishment of a vital records program including back up and offsite storage

_____ Liaison with senior management during the recovery process

_____ Crisis communication plan and procedures, including designated spokespersons

_____ Continuity of management

_____ Continuity with other existing crisis management plans and emergency response procedures

_____ Identification and use of salvage/restoration companies

_____ Training and orientation for new employees

_____ Procedures to maintain and update the plan on a periodic basis

_____ Testing program

_____ Procedures to control the distribution and security of the plan document

_____ Inclusion of recovery planning considerations as part of the organization's strategic planning and new product/service development procedures

Scenarios describe simulated emergency situations, including the overall sequence of events, details, and timing of the specific activities. A scenario provides a set of problems that must be dealt with through the procedures outlined in the emergency management plan. In general, a unique scenario should be developed for each drill/exercise to keep the individuals involved from anticipating events and to ensure a valid test of the participants. For exercises, a unique scenario is required; however, since drills are primarily training activities, the re-use of scenarios depends upon the training objectives.

Scenario development is an interactive process involving several steps. The initial step is development of general scenario guidelines as part of the planning process. These guidelines should address issues of exercise scope and duration, participants, objectives, administrative and logistic considerations, and operational or technical constraints. The following guidance may be used for developing scenario content:

- **Satisfy Test Objectives**: Scenario design must allow all test objectives to be satisfied. Both the overall scenario concept and the individual specific scenario events must accommodate the objectives.

- **Realism**: Scenario information, data, and evidence should be presented as it would be found, measured, or indicated, with a maximum of realism. It should conform, as closely as possible, to actual site conditions, and be based on events that could happen at the facility.

  - For authenticity, and wherever possible, a scenario should take into account and accurately reflect the mission of the facility; its layout, geography, and local environment; the nature of its personnel; its accepted threat levels and potential vulnerabilities; its operating and security plans and procedures; and other considerations, such as props and other visuals should be used. For example, if emergency procedures call for the use of protective equipment or clothing, the actual protective equipment and clothing should be used during the event.

- Scenarios should be designed to conform with the time constraints of the performance tests, and should not be compressed or expanded to fit the length of time scheduled for the test.

The next step is the actual development of a scenario outline, incorporating understanding of the facility environment, the test objectives, and the requirements of the emergency management plan. It should be a sequential listing of the key operational, technical, and logistic events comprising the scenario and the approximate timing of these events. Subsequent steps involve refining the timeline of key events, developing the detailed scenario information, and preparing the specific exercise messages and data. The basic steps in scenario development are:

- **Develop Initial Structure of the Scenario**. Initial ideas must be generated to form the basic structure of the scenario. This can be accomplished by:

  - identifying the emergency situation and the circumstances that caused it

  - identifying the location(s) of the event(s)

  - identifying the subsequent actions and other events that will require additional emergency management functions.

- **Confirm Credibility of the Event**. The potential scenario events developed must be evaluated to confirm that such events could actually take place as postulated. This can be accomplished by:

  - physical inspection of the areas involved in the scenario

  - discussions with facility employees.

- **Ensure Control of the Scenario**. The scenario must be examined for any "fatal flaws" that would result in loss of control of the scenario. For example, there should be no possibility that the Emergency Operations Center could order a single action early in the scenario that would satisfactorily resolve the problem and end the test. The scenario must be constructed so that the test coordinator has ultimate control over the pace of activity and the duration of the test.

- **Ensure Comprehensiveness of Scenario.** The scenario must provide sufficient input to the emergency management team to allow those managing the problem to make decisions. It must provide sufficient background information and other enhancements such as:

  - photographs and personnel/medical records of employees who are involved in the scenario

  - police/intelligence reports pertaining to the situation

  - meteorological data

  - written or taped communications that pertain to the situation

  - videotaped simulated news reports about the emergency

NOTES

NOTES

NOTES

NOTES